G000096082

Mining the Moon
Bootstrapping Space Industry
3rd edition

David A. Dietzler

Copyright-©-2020-by David A. Dietzler.-All-Rights-Reserved

Table of Contents

Preface

This book is the result of years of web surfing and reading about space, science and technology. Hopefully the reader will find inspiration in this work. My approach has been more of a "how to settle space" than a why settle space. It should convince readers that space can be conquered with foreseeable technology and that it can be done profitably.

I wish to thank the editor of the *Moon Miner's Manifesto,* Peter Kokh, for his years of hard work and writing that has informed and inspired me. He originated many of the ideas that are presented in this book. Peter also took me seriously enough to publish many articles that I wrote over the years. If it wasn't for that I probably would have forgotten all about space exploration, industry and settlement a long time ago.

I also wish to thank Mark Rode for his 3D CAD diagrams and untold hours of technical discussions over the past decade or two. I have learned much from Mr. Rode.

I also have to thank Bryce Meyer for his suggestions and ideas regarding life support systems. His work is most enlightening.

I want to thank members of the local National Space Society chapter, the St. Louis Space Frontier, and their dedication to making the chapter do great things from hosting conferences to giving presentations at libraries and the Cortex Innovation Center in St. Louis, Mo.

David A. Dietzler
2020 A.D.

Introduction

Space Resource Basics

Since the 1970s scientists, engineers and space exploration advocates inspired by Dr. Gerard K. O'Neill have been studying the use of space resources, particularly lunar resources, for the creation of space habitat and solar power satellite construction. Space habitat, colonies or more preferably settlements, with masses of millions of tons and space solar power satellites (SSPSs) miles in diameter weighing tens of thousands of tons were envisioned. More recent thinkers have designed lower mass SSPSs and proposed the use of automation instead of thousands of humans in space to build them.[1] The key to success in such endeavors was the Moon mining base, lunar mass driver launcher and mass catcher system.

Rocketing huge tonnages of materials into space from Earth's surface to build SSPSs even with partly reusable SpaceX and Blue Origin rockets would be cost prohibitive. Instead of using rockets, the way to get materials into space would be to launch them from the Moon with a mass driver-sort of an electromagnetic cannon. Expensive rockets that are either discarded or recovered and refurbished for a substantial price would not be used. The mass driver would only need cheap electricity and it could launch anywhere from several hundred thousand to several million tons of material into space every year. Such an enterprise could be engaged in because the Moon has no atmosphere and its escape velocity is much lower than Earth's. It only takes 1/21 as much energy to break free from the Moon's gravity than it does the Earth's. Since the Moon has no atmosphere it is possible to launch materials at several thousand miles per hour while the same attempt on Earth would result in a payload that burned up due to air friction. This material would consist of some refined metal along with lunar regolith which is rich in oxygen, silicon and numerous metals. Spacecraft called mass catchers located at the Earth Moon Lagrange point 2 (EML2) would receive these payloads and haul the material to stations at other Lagrange points or in Earth orbit where it would be transformed into SSPSs.

Lagrange points are regions of gravitational stability near the Moon and at sixty degree angles in lunar orbit. Mass catchers would consist of Kevlar bags several hundred feet in diameter that use rotary pellet launchers for station keeping and propulsion. Loads of regolith weighing about 40 kilograms moving at 200 m/s would impact the mass catcher at EML2 every second. About 5% of the caught mass would be formed into pellets and ejected at 4000 m/s with rotary pellet launchers to counteract the impacts of material.[2] Ablation propulsion with lasers or electron beams is an alternative to pellet launchers. While some have balked at the idea of launching millions of tons of material with a lunar mass driver and crashing this matter into the mass catcher, the situation is more like a constant drip, drip, drip than a tsunami. It should be possible to keep the mass catcher in its place at EML2 until it fills up then hauls its load to a refining and construction station in GEO. Some of the caught material will be sacrificed for propulsion. Ablation propulsion might use the material more efficiently than rotary pellet launchers.

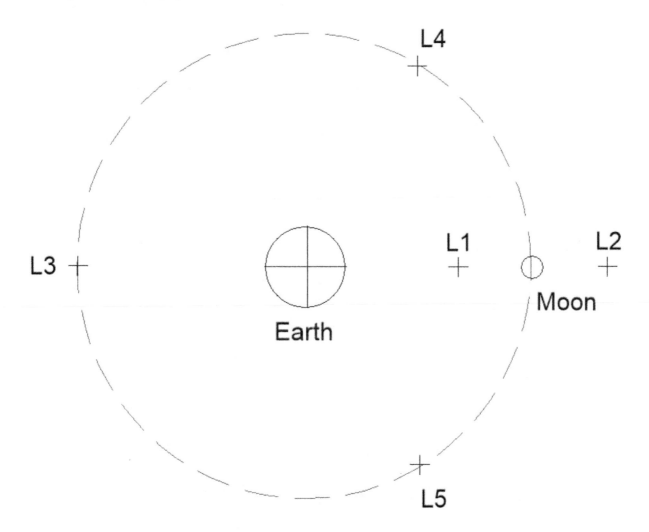

Figure 1. Earth-Moon Lagrange points. L1 and L2 are roughly 60,000 km from the Moon. L4 and L5 are equidistant from Earth and Moon.

The whole system would be more elegant than byzantine in this author's opinion and it is based on existing technology for the mass driver like mag-lev railway propulsion. Even if the SpaceX Super Heavy rocket launches payload to low Earth orbit (LEO) for $67/lb. and then for the cost of some more payload in the form of propellant to move the stuff to geosynchronus equatorial orbit (GEO), this cannot compete with the price of a few dollars per pound of material from the Moon. The only barrier is the high cost of the initial investment in a Moon mining base. If only the machines needed to make machines to make machines and other pieces of equipment with lunar materials are sent to the Moon the cost of the Moon mining base can be slashed. This is called *bootstrapping* and the use of local resources is called In-Situ Resource Utilization or *ISRU.*

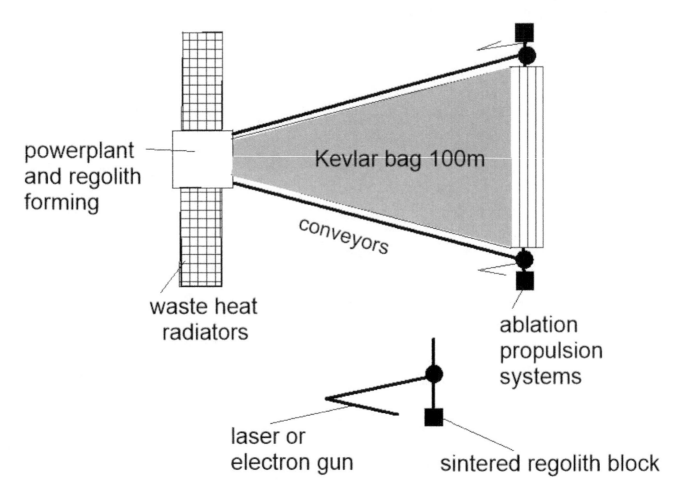

Fig. 2 Mass Catcher with ablation propulsion. Lasers or electron beams vaporize blocks of material to form exhaust jet.

Basics of Rocketry

A basic understanding of rocketry is helpful. The whole field of rocketry is dominated by the tyranny of the rocket equation. The initial mass of a rocket (Mi) when it is fueled and loaded divided by the final mass of the rocket (Mf) after it has expelled all its propellant or reaction mass is called the mass ratio. The natural logarithm of the mass ratio times the exhaust velocity of the expelled mass equals the final velocity or "delta V" of the rocket.

$V = c \log_e (Mi/Mf)$ or $(Mi/Mf) = e^{(v/c)}$ c = exhaust velocity v = rocket velocity

The higher the mass ratio (Mi/Mf) and the higher the exhaust velocity the faster the rocket. At a mass ratio of 2.718 to 1 the rocket will travel at a speed equal to its exhaust velocity. To go twice as fast a mass ratio of 7.4 to 1 is needed and to go three times as fast a MR of 20 is needed. Increases in speed require huge increases in propellant or reaction mass loads. That's the tyranny of the rocket equation. It gets worse. To increase the propellant mass larger fuel tanks are needed and that increases the mass final and requires a reduction of payload mass. A larger rocket carrying more propellant is heavier so it needs more thrust from its engines to get off the ground. To make matters even worse the rocket will endure air friction and gravitational deceleration as it climbs up from the Earth to LEO. When these aerodynamic and gravitational losses are accounted for a rocket needs to reach a theoretical speed of 9 km/sec to 10 km/sec rather than the orbital speed of 7.8 km/sec to reach LEO.

The bright side is that v is directly related to c. If the exhaust velocity is doubled the delta velocity is doubled. Rocket engineers have achieved high exhaust velocities with propellants like hydrogen and oxygen and they have built rockets with low mass fuel tanks using thin sheets of metal. They have also developed high thrust rocket engines. Staging is also used. Liquid fueled chemical rockets because of their high thrusts are just fine for reaching LEO; however, at least two stages are required and once in orbit the rockets' fuel and oxidizer is gone. If you want to go to the Moon or another planet in the solar system you need at least three stages. If it was possible to refuel in orbit the upper stage could travel to other worlds, but how do you refuel? Propellant could be launched up from Earth at great cost. Better yet, propellant could come from the Moon or near Earth objects (NEOs). That would require a large initial investment in Moon and/or NEO mining infrastructure. This would probably be worth it for the future of mankind and any other Earthly life forms that we might take with us into space.

Chemical rockets have reached the practical limit of exhaust velocity with liquid hydrogen (LH_2) and liquid oxygen (LOX) propellants. Nuclear rockets with LH_2 for reaction mass could go faster, but nuclear rockets present problems. Electrical propulsion like ion drive or VASIMR (Variable Specific Impulse Magnetoplasmic Rocket) offers much higher exhaust velocities but much lower thrust. Electrical propulsion can't get off the ground but it can accelerate spacecraft slowly in outer space to very high speeds. Energy can come from solar or nuclear sources. Since electric drives offer such high exhaust velocities either the rocket can reach much higher speeds with a given mass ratio or it can have a much lower mass ratio and much lower reaction mass load than a chemical rocket for the same delta V. If your reaction mass is coming up from Earth at great cost or it is coming from the Moon or a NEO you can save money by using electric propulsion and a smaller load of reaction mass. The only problem is that low thrust means long slow accelerations therefore it might take months to reach lunar orbit when chemical propulsion could get you there in days. This won't matter too much with unmanned vehicles. Electric propulsion can't land on the Moon or most other bodies in the solar system because of its low thrust so chemically propelled landers are still needed.

The way to do things seems to be the use of chemical and electrical propulsion together. If the Moon and/or NEOs can supply relatively inexpensive propellant, spacecraft with high thrust chemical boosters could leave LEO and run up to escape velocity then activate electric drives and accelerate slowly for days even weeks and reach high speeds. This hybrid spacecraft could reach Mars or any other destination in the solar system faster than chemical or electrical propulsion alone. The infrastructure for building SSPSs could also supply propellant in the form of hydrogen, oxygen and combustible metal powders for deep space voyaging.

VASIMR is interesting because it can vary its specific impulse. Specific impulse abbreviated Isp is a rating of a rocket's efficiency. It is directly related to exhaust velocity. The Isp times the acceleration of gravity, 9.8 m/s^2 gives the exhaust velocity, c in the rocket equation, in meters per second. So a rocket with an Isp of 400 seconds has twice the exhaust velocity of a rocket with an Isp of 200 seconds and will go twice as fast with the same mass ratio.

Thrust is directly related to mass flow rate through the engine. If you can double the amount of propellant mass through the engine per second you can double the thrust. Thrust is directly related to Isp. Double the Isp and you double the thrust. With electric drives you have high Isp but low mass flow rates, so you get low thrusts. The power required for a rocket engine goes up to the square of the Isp. Double the Isp and you need four times as much power. This would double the thrust, but if you keep thrust the same or one half that which you get with the doubled Isp by reducing the mass flow, you only need twice as much power. The Isp is directly related to the power required for the same thrust.[3] Keep thrust constant by reducing mass flow and doubling the Isp only doubles the power requirement.

With VASIMR, it is possible to get a higher thrust at a lower Isp and accelerate to escape velocity. As the rocket speeds up it can lower its thrust by decreasing mass flow through the engine and increase its Isp by adding more power directly with increase in Isp. In this way it can "shift gears." Ultimately, with the higher exhaust velocity or Isp it can reach high speeds and get to Mars in about 39 days.[4] This would require a lightweight high power reactor.

In conclusion, resources from the Moon and eventually NEOs, combinations of chemical and electric propulsion, lightweight nuclear and solar power systems and low cost reusable rockets to LEO can open the doors to the solar system. Lunar and asteroid resources can also supply more than propellant. They can supply oxygen, water and food once space farms are established.

Life Support Basics

Rockets of all kinds can get us into space. Life support systems (LSS) will make it possible for us to stay in space for long periods of time. Early LSS for spaceships and habitat will use supplies of oxygen and chemical CO_2 absorbents. It might also be possible to freeze CO_2 and other pollutants out of the interior air with cryogenic systems. Extra oxygen could be extracted from lunar rocks and regolith. Eventually, space farms and green plants will recycle the CO_2 and generate oxygen. This is called Closed Ecological Life Support Systems or CELSS.

Water can be recycled by condensing humidity from the air and filtering, distilling, purifying and sterilizing waste water from sinks, showers, toilets, etc. Mechanical systems can do this until CELSS is established and biological systems for recycling water are in action. Food is mostly water so dehydrated and freeze-dried foods will be rehydrated with the same water over and over again. Whole food would be very bulky and water supplies to sustain crews for months at a time would be far too massive and expensive to ship into space.

Ultimately, it will make more sense to grow food in space instead of ship it into space even if it is dried out. Space farms must be established early on at manned bases so that the first harvests can be made within a few months. Fresh food will be much better for morale than dried foods. Crops will require controlled temperatures. So will humans. Habitat will require thermal insulation plus heating and cooling systems. Cosmic ray shielding will also be necessary. This would probably consist of several meters of regolith to cover the habitat. That would also serve as a barrier to micrometeorites.

Chapter 1: Why Mine the Moon?

Why return to the Moon? Mostly to build up industry to the point at which millions of tons of materials can be launched into space inexpensively every year for the production of:

1) Energy from solar power satellites. Earth will need about 60 TW by the mid 21st century based on present rates of growth; however, that's primary thermal energy and only a third really gets used. The rest is waste heat. If we go all electrical we will need 20 TW electrical to power civilization. This would require one thousand gigantic powersats rated at 20GWe each in addition to receiving antennas (rectennas) on the ground and a whole new heavier power grid. This is an enormous challenge. Chances are that much energy in the future will come from ground based solar, winds, hydro, geothermal, waves, tides, biofuels, nuclear fission and some fossil fuels. It's more likely that a smaller number of space based solar power satellites rated at perhaps 5GWe to 10GWe will make a significant contribution to the clean energy mix and earn large profits for their owners. Helium 3 mining for fusion fuel is another possibility.

2) Large space stations for tourism, research, and astronomy.

3) Large GEO telecommunications space stations. As global data traffic and telecommunication grows into the future the number of slots in geosynchronus Earth orbit (GEO) for satellites will be used up. Powersats would cause even more interference. Communication satellites will have to be replaced by orbital platforms that have numerous transponders and antenna arrays. These would have to operate on frequencies different from those used by powersats and/or make use of laser beams.

4) Possibly micro-gravity factories that make things that can only be made in space like exotic alloys, protein crystals, pharmaceuticals, etc. Industries that require high vacuum for clean rooms to make micro-electronics and other products might find a place in space or on the Moon.

5) Space shipyards to build ships for flights to the Moon or Mars and robotic mining ships to asteroids as well as propellant for these ships. Lunar liquid oxygen (LUNOX), hydrogen from lunar polar ices or silane, raw regolith for mass driver propulsion, sodium and perhaps magnesium for ion drives could be used to fuel ships. There would be plenty of lunar aluminum, magnesium and titanium alloys to build these ships.

6) Space tourism in orbit, on the Moon, even Mars.

7) Lunar observatories and huge telescopes in space that can actually see exoplanets well enough to see cloud patterns and continents of the nearer exoplanets and spectroscopically study them. Life could be detected by studying the gases of their atmospheres or seeing the spectrum of chlorophyll.

8) Exploratory robotic and manned ships to Europa, Titan, Enceladus, Mercury and beyond.

9) Orbital space settlements built of lunar and asteroidal materials. These could be large enough to house thousands of people and farms to provide food for workers and tourists in space.

10) The construction of defensive and law enforcement bases and stations.

11) Consider the spin offs as all this technology is developed as well as the infinite supply of clean energy to the Earth. A lot of AI development will be required and the hardware and software for this could find all sorts of unpredictable applications.
.
12) With lunar materials large fleets of ships with thousands of settlers could be sent to Mars and the planet could be terraformed or if you prefer areoformed or rejuvenated. What we learn on Mars will be applied in the more distant future when we travel to other star systems and find Mars-like planets. There could be other worlds where life has never evolved but the right conditions exist for the planting of life on those worlds. The things we learn about mining asteroids and building space settlements--veritable cities in space, will be applied in other star systems which most probably have asteroids and comets also.

13) Let's not forget that filmmakers will have so many opportunities that several cable channels could be filled with shows about what's going on up on the High Frontier and that these new frontiers will inspire untold generations of new artists, musicians, poets, writers, sculptors and other creative types. There could be televised athletic competitions in low G, weightless dancers, circuses, and more broadcast live.

14) HOPE for a long lasting future for the human race and no more end of the world scenarios. The development of trade between the Moon, Mars, asteroids, space colonies, settlements on other worlds and subsequently a larger economy that offers new business opportunities and a chance for entrepreneurs to tap new markets. New generations of leaders will develop in space.

Chapter 2: Lunar Money Matters

How much would it cost to build a bootstrapping base on the Moon for supplying materials for solar power satellite construction and other projects on the Moon and in orbit? We can make some estimates. A corporation would have to be formed that was capitalized by private multi-billion dollar fortunes, large international companies and by the sale of stocks and bonds. Research and development of all hardware and software would be required. That alone might cost one or two billion dollars.

If the initial batch of equipment, the "lunar industrial seed," weighs in at 1000 tons and a Falcon Heavy can put 53 tons in LEO for $90 million, the cost of orbiting the seed would be $1.7 billion. By comparison, the Space Shuttle could orbit 30 tons for $450 million to $1.6 billion.[5] The cost of the actual equipment might be one or two billion dollars. So now we are at $3.7 billion to $5.7 billion.

Transporting the "seed" to the Moon will not be cheap. Solar electric cargo tugs will have to be orbited and they will need propellant although not a lot compared to chemical rockets. We can estimate that the tugs and their hydrogen, argon or xenon propellant loads will total about 300 tons or 30% as much as the payload mass. Orbiting those would cost about $500 million. Landers and space storable hypergolic propellant to move the cargo from low lunar orbit (LLO) to the lunar surface would be needed. That could add up to another 1500 tons to LEO that then has to be moved to LLO. It would cost about $2.5 billion to put this in orbit. We are now looking at $6.7 billion to $8.7 billion.

This estimate did not come out of thin air. We can use the rocket equation to determine that with hypergolic propellant fueled lander rockets with specific impulses of about 315 seconds and a delta V of 1.8 km/sec to descend from LLO and disposable one way landers that weigh about 20% as much as the 1000 ton lunar industrial seed, about 950 tons of propellant is needed. To move this mass of 1150 tons to LLO we can estimate that 30% of that for solar electric cargo tugs will bring the grand total for tugs, landers and propellant for landing 1000 tons of actual cargo is just about 1500 tons.

Landers without propellant onboard that weigh only 20% as much as their payloads seems like a stretch. Unlike a manned lunar lander that has two even three stages for descent, ascent and orbital rendezvous these landers would have one stage, travel one way and be expendable. They would consist merely of fuel and oxidizer tanks made of lightweight 8000 series lithium-aluminum alloy or composite materials along with a small rocket engine. They would not have landing gear but only airbags. Upon touchdown the landers would probably be crushed by the heavy cargo modules they are attached to. They could then be scavenged for materials. The cargo modules containing the actual payloads of machines, supplies and any packing materials would also be scavenged. Nothing can go to waste in the austere environment of the Moon.

There would also have to be ground stations for tracking and control and that would add cost. A wild guess is that about $10 billion would be involved. There might be ways to reduce costs. An exhaustive study could be done to determine exactly how much cargo for the lunar industrial seed would really be needed to build up a full-fledged mining base. Perhaps only a few hundred tons of cargo rather than a thousand tons as I have speculated would be needed. If NASA builds a base in the south polar regions of the Moon and successfully mines ice, that ice could be used to make propellant for reusable landers that shuttle cargoes down from LLO to the base site. The best place to put a mass driver launcher on the Moon seems to be at 33.1 degrees east on the lunar equator.[6] If NASA can supply lander propellant to a commercial mining and mass driver base on the equator for a reasonable price over a billion dollars might be saved.

How much money can the base earn? At $2 per kilogram launched by mass driver with mass catchers expelling about 500,000 tons of reaction mass for delivering 600,000 metric tons of lunar material annually, that's $1.2 billion per year. It would take about 8.33 years to break even if $10 billion is spent. An extensive amount of space industry would have to exist. Solar power satellites are the only way to make a business case for lunar industrialization. When we consider that a 5GWe space solar power satellite selling electricity at 2 cts./ kWhr could earn revenues of $867 million annually and $2.19 billion annually at 5 cts./kWhr, everything appears to be more lucrative. Wholesale electricity is sold for about 5 cts./kWhr today and retail goes for 10 to 11 cts./kWhr. In some places, like Germany and California, retail electricity is even more expensive. Space solar power satellites could earn substantial profits selling power to developed nations and still earn plenty selling cheap electricity to developing nations. Global living standards could be improved without resorting to large scale burning of cheap fossil fuels in developing nations.

Space industry, energy and settlement will not be the boondoggle some people may think it is when cheap clean power is delivered to the Earth. The real travesty would be continuing to burn coal for electricity in ever larger amounts with no concern about climate change, acid rain and other adverse effects on the environment. Moon mining and SSPSs are not the end of the story. Tourism, space manufacturing, space real estate and other enterprises could lead to substantial financial gains. Wealthy space condominium and time share owners will not be the only people who enjoy the final frontier. Many people will go to work in space and some of them could become permanent citizens of space. Eventually the price of space travel will fall and a mass market will evolve. No doubt there will be plenty of pioneers who don't care about NW Canada, Siberia or Antarctica who emigrate out into the solar system.

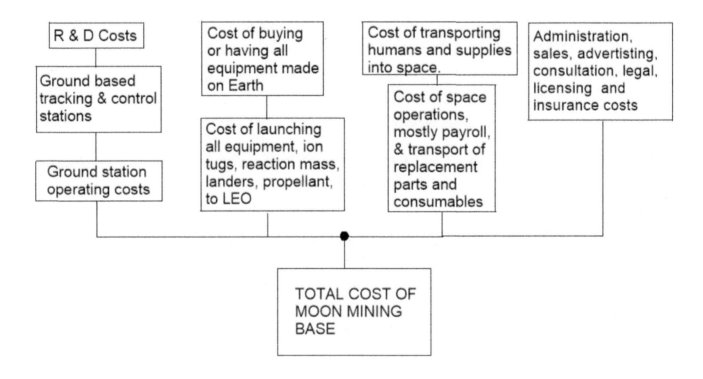

. Fig. 3 Flow chart of cost items contributing to price of a lunar industrial base

Chapter 3: Lunar Resources

The Riches of the Moon

The Moon, sitting in a gravity well only 1/21 as deep as Earth's, could supply millions of tons of oxygen, raw regolith, basalt, ceramics, glass, silicon, iron, calcium, aluminum, magnesium, titanium, sodium, sulfur, potassium, manganese, chromium, volatiles and more annually for construction in outer space. Uranium, thorium and rare earth elements exist on the Moon and the possibility of undiscovered igneous ore bodies of magmatic or volcanic origin sparks the imagination. Ice in permanently shadowed polar craters could supply water, carbon monoxide, carbon dioxide, ammonia and possibly other substances. Rockets would be used to transport mining equipment, metal extraction and materials processing devices and manufacturing machines like 3D printers, large engine lathes and assembly robots to the Moon. The Moon base would be built up with local resources. This is called *in-situ resource utilization or ISRU*. Machinery, habitat, vehicles and more will be made on the Moon using an initial stock of carefully selected machines to expand the base and replicate machinery. This is called *bootstrapping*. Once the base is built up an electromagnetic mass driver launcher will shoot payloads of regolith and other materials into space for construction. This will not be the end of the Moon base. Lunar industry will continue to expand and multiple settlements will be established on the Moon connected by dirt roads and railways.

Besides physical materials, the Moon offers a number of unique resources. There is free vacuum for establishing super clean environments, purifying and evaporating materials, and making things without contamination from atmospheric moisture, oxygen and nitrogen. Rust, corrosion, storms, lightning strikes, floods, forest fires and other Earthly woes are non-existent on the Moon. Low gravity makes it possible to move heavy loads around without as much manpower or horsepower. Loads on machinery and bearings will be reduced. Solar energy is available reliably for two weeks at a time, never obscured by clouds, at lower latitudes and in polar regions it is available up to 90% of the time. Solar energy can be used to generate electricity or it can be concentrated with parabolic mirrors to produce high temperatures for melting materials. Super cold temperatures can be obtained with shielded space radiators exposed only to the 4K temperature of outer space. This makes it possible to liquefy oxygen, hydrogen, helium and other gases with ease.

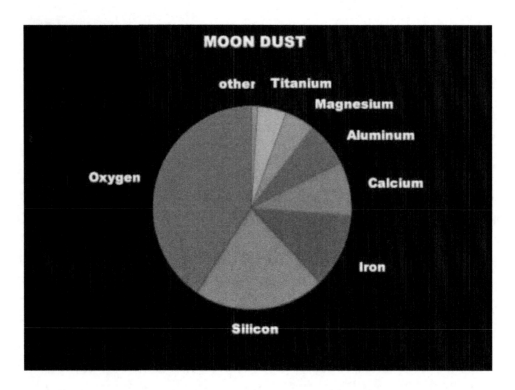

Fig. 4 Lunar soil or Moon dust, technically called regolith, contains many useful elements for survival and industry on the Moon and in outer space.

Lunar soil, regolith, or Moon dust is mostly oxygen and silicon along with iron, calcium, aluminum, magnesium, titanium and traces of chromium, manganese, sodium, potassium, phosphorus, sulfur and tiny amounts of many other elements. It also contains traces of hydrogen, helium, methane, CO, CO_2 and nitrogen implanted by the solar wind that can be extracted by mining millions of tons of regolith and heating it to about 700 C. Numerous processes for extracting and separating these elements have been described. Ideally, it should be possible to heat regolith until it decomposes and separate the elements with something similar to a mass spectrometer. Dr. Peter Schubert of Purdue University has designed a device that heats and decomposes regolith into oxygen and other elements like iron, aluminum, titanium and silicon. These are all separated in what he calls an "All Isotope Separator" that resembles a mass spectrometer or calutron for enriching uranium.[7] A slag of calcium and magnesium oxides forms too. Magnesium can be extracted by heating this slag mixture with silicon at about 1200 C. under reduced pressure. Magnesium evaporates and is condensed. The calcium silicate slag remaining can be broken up and used as aggregate in concrete or it can be processed to get calcium compounds and even pure calcium metal.

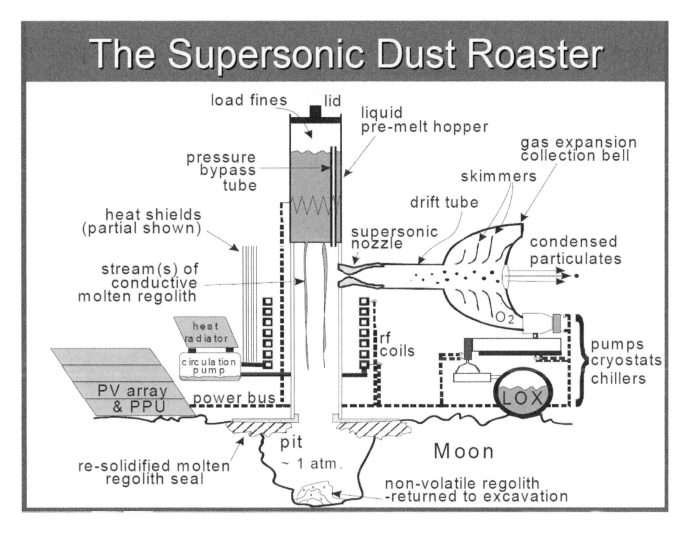

Fig. 5 Dr. Schubert's device for decomposing regolith with intense heat for oxygen and other elements.

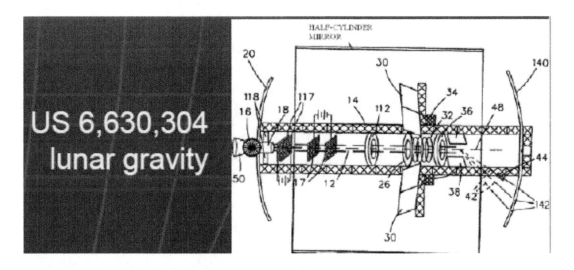

Fig. 6 An All-Isotope-Separator for operation in low lunar gravity.

Numerous processes for extracting oxygen and metals from lunar regolith have been proposed. Most of them use chemical reagents like hydrofluoric acid, fluorine, chlorine, lithium and other substances not common on the Moon. These chemicals will corrode things, leak and become contaminated over time. They will have to be replenished with imports from Earth at great cost. A process that doesn't require water and/or lots of corrosive and imported chemicals is desired. The Supersonic Dust Roaster and All Isotope Separator (SDR-AIS) fits the bill.

The SDR-AIS is based on complex laws of physics and is rather sophisticated. However, its operation is simple. Loads of regolith go in. Oxygen and metals come out. It is also much simpler than the hydrofluoric acid leach process studied by NASA.[8]

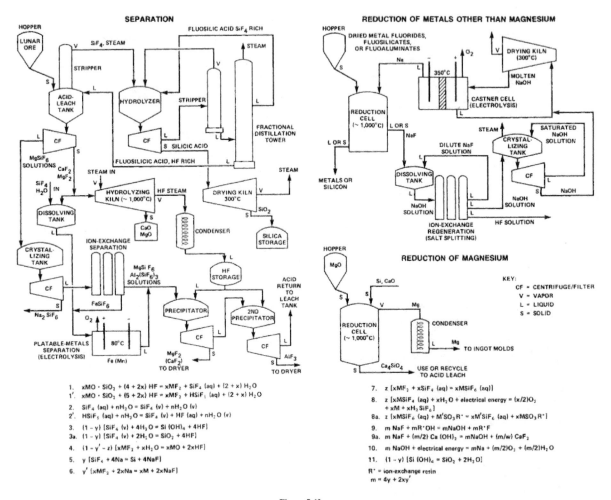

Figure 5.41. -
Flowsheet and process equations for the HF acid-leach process

Fig. 7 HF acid leach process from NASA's Advanced Automation for Space Missions study done in 1981.

This system consists of 34 component leaching tanks, distillation towers, centrifuges, precipitators, electrolysis cells, etc. There are 111 pipe connection points with at least one valve and control mechanism. Today's computers could easily control all this but the complexity equates with lower reliability and higher costs.

Simple Materials

Basaltic mare regolith can be excavated, pressed into forms and sintered with heat from solar or electrical furnaces, or melted and cast to make numerous basalt items. Highland regolith can be sintered or cast to make glassy products with a melting point of about 1500 C. Meteoric iron fines can be magnetically extracted from large amounts of regolith just about anywhere on the Moon. After some grinding, screening and a second magnetic separation, 99% pure iron-nickel particles can be obtained.[9] This can be refined further with carbon monoxide to extract nickel. Another possibility is the use of a device similar to a mass spectrometer to separate iron, nickel, cobalt, gallium, germanium and platinum group metals from this meteoric material. It will also be possible to melt and cast the iron fines in sand molds. It should be possible to run these particles in 3D printers after sizing them to make iron-nickel items for low stress applications. It will also be possible to melt and cast the iron into rods, pack them in carbon powder, and get them red hot for a few days in solar or electrical furnaces to convert the iron into strong steel. So far, so good. Mining the Moon for these materials doesn't require any water, acids, halogens or other substances rare or practically non-existent on the Moon.

Valuable Light Elements

Light elements like hydrogen, nitrogen and carbon are of great use. Hydrogen can be combined with oxygen to make water or it can be used as rocket fuel. It can also be reacted with carbon to make chemicals and plastics. Nitrogen can be used to make synthetic materials and it is needed for agriculture. Carbon is needed for agriculture and it is the backbone of organic chemicals and plastics.

Heating regolith to 700 C. will release hydrogen, helium and nitrogen. Carbon will react with oxygen in the regolith to form CO and CO_2 gases. Hydrogen will react with carbon to form methane and it will react with oxides to form water. These gases could be separated by fractional liquefication. At the University of Wisconsin in Madison, scientists have designed a machine that could dig up and heat regolith from a square kilometer of lunar surface to a depth of three meters in a year's time. They predict harvesting of substantial quantities of volatiles.[10]

Materials Harvested per Year in Tonnes

Element/Compound	Mass
water	109
nitrogen	16.5
CO_2	56
hydrogen	201
helium 4	102
methane (CH_4)	53
CO	63
helium 3	33 kg.

Enough hydrogen could be harvested to make 1800 metric tons of water when combined with oxygen extracted from regolith. This could support a population of several hundred persons on the Moon. The CO_2 molecules are very stable and require temperatures of 3000 C.+ to decompose at extremely low pressure.[11] The CO molecules have an even higher decomposition temperature and they can react with each other to form CO_2 and carbon.[12] It will require less energy to react these carbon oxides with hydrogen to make methane which can be decomposed at 900 C. to get pure carbon and recover hydrogen. The total amount of carbon from the masses of CO, CO_2 and CH_4 above will be 82 tons at 100% recovery. That's enough to make lots of steel and some plastics too. Chances are less carbon than that will be obtained, but the CO and CO_2 remaining will have uses.

Polar ices may also serve as a source of water and thus hydrogen and oxygen for fuel cell reactants and rocket propellants. LCROSS data showed that the ice contains ammonia (NH_3), methane (CH_4), carbon monoxide and carbon dioxide. The CO_2 will be needed for plant growth in hydroponic farms. Hydrogen and CO can be combined to make many different organic chemicals. Ammonia can provide nitrogen for atmospheres and fertilizer.

Resources for the Future

Excavators, transports, solar power plants, regolith refining equipment, manufacturing machines, vehicles, inflatable habitat, superconducting wire for the mass driver launcher and robots will be landed on the Moon. Some pieces of equipment will be made partly on the Moon. Eventually all things including water and food will be produced on the Moon.

Lots of cargo will be required to do this job. It will be necessary to figure out how to make all this on the Moon from an initial stock of devices or a "lunar industrial seed" that might have a mass of several hundred to several thousand tons. Energy requirements and times for each step of the process must be calculated; a job for chemical and industrial engineers. The economics of all processes must be determined by doing research on the ground in vacuum chambers that simulate the lunar environment, at lunar analog research bases and at a Lunar Industrial Research Park on the Moon where hard facts are determined.

Naturally, many solar panels will be needed to power this equipment. The 3D printers, casting foundries and machine shops where we replicate this regolith refinery equipment will not be limited to the expansion of the refinery. They will also use lunar materials to crank out all the other items needed for life in space including bathroom and kitchen fixtures, plumbing systems, everyday items like dishes and bottles, robot parts, vehicle parts and complete vehicles, excavator parts and complete excavators, spacecraft and more 3D printers and machine tools.

At some point in time, growth of the lunar mining, refining and manufacturing complex will become exponential. The first base will produce enough stuff to build a second base, then a third and fourth base, then eight bases, sixteen bases, etc. We will need dirt roads and eventually railways to connect the bases. A vast amount of solar power generation and distribution infrastructure will also be built.

Chapter 4: Bootstrapping

1. Location

The first thing that must be done before the construction of a lunar industrial facility is determining the best place to locate it. Polar locations offer craters containing ice and prolonged sunshine 80 to 90% of the time. However, the best place to locate a mass driver launcher is on the lunar equator at 33.1 degrees East longitude. While NASA might build a research base in the South polar region, industry should locate in a place suitable for mining and mass driver operation.

2. Site Preparation

Once a site is selected it needs to be prepared. The site must be leveled and small craters must be filled in. Boulders must be dynamited and the rocks pushed aside. Robots that can drill holes in boulders and place explosives will also be needed. Logically, all the robots will retreat to extreme range when boulders are blasted.

Markers must be placed to indicate the locations of solar panel farms, landing pads, roads, walkways, a warehouse, a pad for production machinery and inflatable habitat modules. Robotic bulldozers and graders will be called for. There must also be a solar panel farm and wiring systems to recharge the batteries in the bulldozers and graders. These machines might be powered by tethers or microwave beams from the solar panel farm. Receiving antennas on the machines will just be low mass wire meshes with some Zener diodes and this will not burden the 'dozers and graders. This would free the machines from the burden of heavy battery or fuel cell packs and the need to shut down and recharge for several hours at a time.

Clearly, the first payloads to the Moon must be solar panels and associated hardware along with several bulldozers and graders. Robots to deploy the solar panels and wiring systems will also be needed. To protect the machines during nightspan it might be desirable to have infrared lamps to keep parked machines warm and power storage systems to energize the lamps. Batteries, flywheels or fuel cells come to mind.

Fuel cell systems will require insulated tanks to store liquid hydrogen and liquid oxygen, plumbing systems, water electrolysis systems, and refrigeration devices to liquefy hydrogen and oxygen. That sounds like a complicated mess when compared to batteries or flywheels; however, there is an important advantage to the use of fuel cells for night span power storage. Fuel cell systems can augment rocket fuel storage facilities with their cryogenic reactant storage tanks and liquification systems.

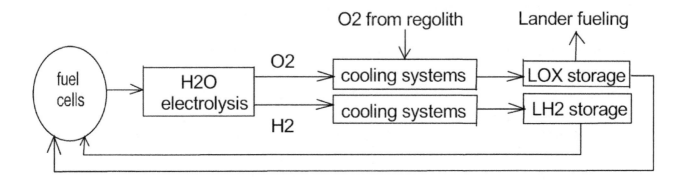

Fig. 8 Fuel cell set up

It should be possible to equip bulldozers with mining shovels so that these can do two jobs instead of just one. Attempting to define the components of a lunar "industrial seed," we can imagine the first payloads to the Moon will include but not be limited to a lunar LOX (LUNOX) production plant, general purpose teleoperated robots, solar panels with supports, motors, reflectors, wiring, switches and invertors, power storage systems, microwave power transmitters or tethers, bulldozers and graders that can also excavate, IR lamps, oxygen generators and pumps to load up lander rockets.

3. Early Development

Once the site is leveled out and large rocks removed, it will be developed. Landing rockets will cause dust to spray all over and dust could damage machinery especially if it gets into bearings. Dust sprays could disaffect solar panels also. Several landing pads will be made. Wheeled robots with microwave generators could sinter or melt the basaltic ground to a depth of several inches (5 to 10 cm) at least. Bulldozers could berm up regolith around the pads. With three pads one rocket could be lifting off while another lands and a third one is waiting for service. Landers or "Moon Shuttles" might have wheels on their landing legs so they can be towed off the pad. The landing pads should be fairly big. A diameter of one hundred meters will allow a large margin of safety if a rocket is a bit off course. The pads would be located about a kilometer away from the habitat so that the chance of a Moon Shuttle rocket going off course and crashing into the habitat and killing everyone is very low. Roads from the landing/launch pads to the habitat and work area will be paved with microwaves.

Lunar workers, their machines and robots will need a nice hard floor made by microwave roasting of the basalt as is done for the landing pads and roads to mount production machines on. Plain old ground is no good. Spacesuited human workers and robots would kick up dust and some machines like power forging hammers would pound or vibrate into the dusty surface. The floor might be thicker than the pads and roads. A large foil or aluminized Mylar parasol to shield machines and workers from the hot sun could be erected and teleoperated robots could work the production machines. Now and then humans will have to go outside in turtleback spacesuits to do some work. There will be microwaved walkways from the habitat modules to the production machine area. There will also be a warehouse consisting of a microwaved pad with a parasol to store cargo containers as they arrive by Moon Shuttle.

In addition to solar panel farms there must be power storage for nightspan not only to keep machines warm with IR lamps but to power lights, radios, computers and mechanical life support systems in habitat modules. A small nuclear generator would help. The microwaved basalt pads will serve as "thermal wadis" and cool slowly after sunset. That will be easier on the machines. Sudden thermal shock can crack metals. Even in polar locations there will be periods of darkness but these will last only a couple of days while in lower latitudes where the mare are darkness will prevail for two weeks out of every month. In the distant future there could be a solar power satellite at Earth-Moon Lagrange point 1 (EML1) and a circumlunar power grid with solar panel farms around the Moon to supply full power at all times.

Secondary payloads to the Moon will include but not be limited to solar panels, wiring systems, power storage and possibly a small nuclear generator. Also needed will be at least two rovers, preferably more, in case one breaks down with microwave generators to make pads and roads. Inflatable habitat with mechanical life support systems and some tanks of oxygen to inflate the habitat will be needed. There should be parasols with support poles to protect workers and equipment from solar heat and to prevent radiation of heat from equipment by night and some running lights, flood lamps and radio antennas. For workers there must be supplies of dehydrated and freeze-dried foods, drinking water and medicines. Clothing, bedding, towels and wash cloths, light weight furniture, recreational supplies (dart board, chess set, playing cards, board games, dice etc.), toiletries and sundry items (toilet paper, toothpaste, brushes, razors, blades, soap, lotion and shaving cream) will be needed on day one. Roll on and pump spray deodorant, cologne, and perfume will have to wait until lots of air cleaning vegetation is cultivated within habitat.

At least this much should be in place before human crews move in and start working. The bulldozers and graders with shovel attachments must cover the inflatable habitat with six meters of regolith for radiation, thermal and micrometeoroid protection. Protection from galactic cosmic rays in free space would require about 11 tons of regolith per square meter of hull to reduce radiation dosage to 20 mSv/yr and 6.6 mGy/yr. On the lunar surface, the Moon blocks out half of this so 9 tons per square meter would be needed.[13] Since bulk regolith is about 1.5 times as dense as water 6 meters of regolith would suffice.[14] Without shielding humans could make only brief sorties on the Moon. It is foreseeable that robots might experience glitches that halt the project and humans become necessary to get things going again. Space workers could land and stay inside their spacecraft for a few days until they get the machines back up and running. Eventually it must be possible for humans to stay on the Moon for extended periods of time and that will require well shielded habitat.

4. Excavating

The bulldozers and their shovels are just the beginning. Massive amounts of regolith must be moved to support a serious Moon mining operation with the goal of building mass drivers, solar power satellites and other constructions in space. A slusher system seems best.[15] This consists of a bucket attached to some steel cables. A winch pulls the bucket through the dust and it picks up a load. The load is lifted and dumped into a truck or an ore car on rails. A second set of cables wrapped around some pylons with pulleys at the edge of the excavation is pulled on by the motorized winch and the empty bucket is dragged out and readied to scoop up another load of regolith. This will be more efficient than making excavators scoop up a load, carry the load and their own weight to the refinery, dump the load and drive back to the hole, and repeat the process. The slusher can work continuously. At first trucks will haul lunar regolith to the refinery on microwaved basalt roads perhaps. Later on, a railway system will be constructed. Cars riding on steel rails will endure much less rolling friction and that will save energy. They won't kick up dust either. When the slusher has dug up a pit and can dig no more it can be relocated. Rail systems can be extended to the new dig site. Slushers will be set up in different locations to dig mare regolith and highland regolith. Mare regolith is richer in iron, magnesium and titanium and can be melted down and cast as is. Highland regolith is richer in calcium and aluminum. These "soils" would be processed differently depending on what substances are desired. The Moon Shuttles will land a few more payloads at this time including slusher systems consisting of cables, buckets, motorized winches, pylons and hauling trucks.

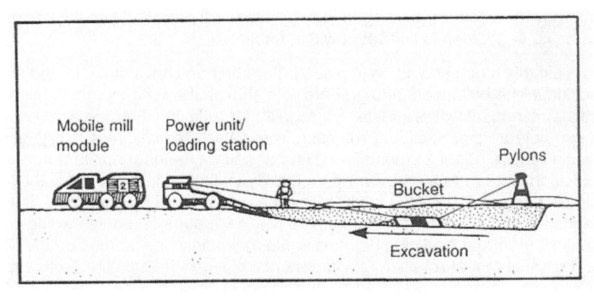

Fig. 9 Side view of slusher mining system. NASA

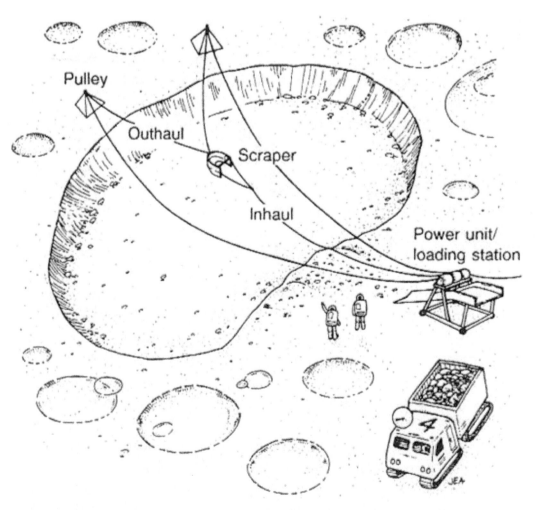

Fig. 10 Open pit mine with slusher system. NASA

5. Materials Production

There are many proposals for the extraction of materials from lunar regolith. Even without complex electrochemical systems for extracting metals there are resources of great value. Mare regolith is basaltic. It can be dug up, melted in a solar or electrical furnace, and crude castings can be made in molds dug into the ground and finer castings can be made in iron molds. It can also be sintered instead of cast. Sintering means that the material is compacted into molds and heated only enough for the edges of its particles to fuse together. This can make worthy items like bricks, blocks, tiles, slabs and rods without as much energy as required by full melting and casting. Basalt can also be melted and drawn through platinum-rhodium bushings to make fibers. A wide variety of things, perhaps even habitat modules and solar power satellite frames, could be made from basalt and glass. Metals will still be needed for vehicles, digging machines, railways, spacecraft, electrical systems, electric motors, etc.

Basalt could be a very important base material. It is harder than steel and abrasion resistant. It is strong in compression but not so strong in tension and it is rather brittle. Uses for basalt include: [16]

Cast basalt

machine base supports (lathes, milling machines), furnace lining for resources extraction operations, large tool beds, crusher jaws, pipes and conduits, conveyor material (pneumatic, hydraulic, sliding), linings for ball, tube or pug mills, flue ducts, ventilators, cyclers, drains, mixers, tanks, electrolyzers, and mineral dressing equipment, tiles and bricks, sidings, expendable ablative hull material (possibly composited with spun basalt), track rails, "railroad" ties, pylons, heavy duty containers for "agricultural" use, radar dish or mirror frames, thermal rods or heat pipes housings, supports and backing for solar collectors

Sintered basalt

nozzles, tubing, wire-drawing dies, ball bearings, wheels, low torque fasteners, studs, furniture and utensils, low load axles, scientific equipment, frames and yokes, light tools, light duty containers and flasks for laboratory use, pump housings, filters/partial plugs

Spun basalt (fibers)

cloth and bedding, resilient shock absorbing pads, acoustic insulation, thermal insulation, insulator for prevention of cold welding of metals, filler in sintered "soil" cement, fine springs, packing material, strainers or filters for industrial or agricultural use, electrical insulation, ropes for cables (with coatings)

More everyday items that could be made of cast, sintered or fiber basalt include plates, dishes, mugs, tea cups, bowls, tea and coffee pots, serving trays, pitchers, decanters, counter tops, kitchen sinks, table tops, table legs, stools, chairs, bars, shelves, bottles, jugs, hand basins, toilets, bath tubs, shower stalls, bidets, planting containers, flower pots, vases, lamps, water pipes and sewer pipes, ash trays, paper weights, candle sticks, aquaculture tanks, floor, ceiling and wall tiles, bricks, blocks, towel racks, clothes racks, shower curtain racks, shower curtain rings, shower curtains from basalt fiber, drapes, cushions of woven fiber stuffed with fiber, mattresses, rugs, statuary, doors, handles and knobs for doors and drawers, picture frames and certainly other things. Small items could be cast or sintered in 3D printed iron molds while large items like toilets and aquaculture tanks could be cast in expendable sand molds bound with sodium silicate.

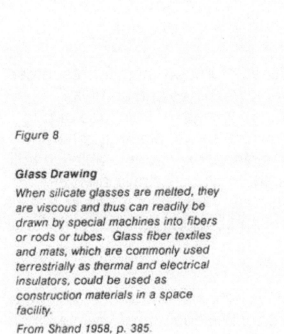

Figure 8

Glass Drawing

When silicate glasses are melted, they are viscous and thus can readily be drawn by special machines into fibers or rods or tubes. Glass fiber textiles and mats, which are commonly used terrestrially as thermal and electrical insulators, could be used as construction materials in a space facility.

From Shand 1958, p. 385.

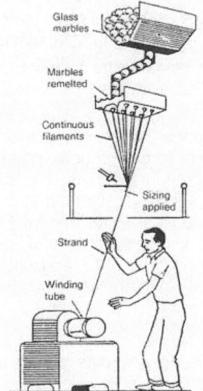

Fig. 11 Basalt fibers can be made the same way glass fibers are made. NASA

Iron molds seem like heavy cargoes to export to the Moon. Perhaps they could be made on the Moon in large numbers. Magma electrolysis yields ferrosilicon and silicate ceramic as well as oxygen.[17] Ceramic blocks could be cast in molds dug in the ground. It might be possible to perform serial magma electrolysis in which case iron could be derived separately from silicon. This iron could be powdered and fed into 3D printers that use electron beams or lasers to fuse metal layer by layer to make all sorts of shapes. If serial magma electrolysis is not possible there is another resource of great value on hand--meteoric iron-nickel fines that are present in regolith all over the Moon at concentrations of a few tenths of a percent by mass. These could be harvested by rovers that have low intensity magnetic separators. This could be the first metal produced on the Moon. The particles are fused with silicates and can be purified by running them through centrifugal grinders to shatter the brittle silicates followed by another magnetic separation. In 1981, Dr. William Agosto projected that this system could produce 552 tons of a 99% pure iron/nickel feedstock annually.[18] After sieving and sizing the powder could be placed in 3D printers to make iron molds of various sizes and shapes for casting and sintering basalt.

Meteoric iron/nickel fines can be used for more than making molds for casting or sintering basalt. They contain 5 to 10% nickel, 0.2% cobalt and traces of germanium, gallium and platinum group metals (PGMs). Iron, nickel and cobalt can be separated by treating the fines with carbon monoxide gas. High temperature vaporization, ionization and electrostatic separation might also be applied. Nickel and PGMs have catalytic properties. Nickel can make steel harder and stronger without making it more brittle. Cobalt can be used for high speed drill bits and cutting tools. It can also stain glass a deep blue. Germanium and gallium can be used in electronics and photocells.

A third way to obtain iron involves the roasting of regolith at 1200 C. in the vacuum to drive off FeO, condensing the iron oxide and reducing it with hot hydrogen or using electrolysis to free up the iron and obtain oxygen.[19] As discussed earlier, a device that thermally decomposes the regolith and separates the elements with a system similar to a mass spectrometer could be used.

Steel could be made on the Moon once iron is available. Carbon from mining for solar wind implanted volatiles and/or polar ice mining can be combined with iron by the simple "crucible" or "blister steel" process. Iron powders, rods or plates would be packed with carbon powder and brought up to red heat (about 1100 C.) in a furnace made of basalt or a ceramic made on the Moon for about a week. The carbon will dissolve into the iron and form steel. The steel and carbon could be magnetically separated and the steel could be homogenized by melting to disperse the carbon evenly throughout the metal. During this melting the steel could be mixed with calcium aluminate flux to remove impurities. The $CaAl_2O_4$

flux could be produced by roasting highland anorthite at 2000 C.[20] Since steel is 0.05% to 1.5% carbon, a little carbon can make a lot of steel. When pure iron won't do, we could make some steel. Since meteoric iron fines contain nickel, some very strong alloy steel could be made from them.

Spacecraft and ground vehicles could be made from aluminum, magnesium and titanium. Bearings, machinery, tools, knives and a few other things will demand steel. Steel cables come to mind but abrasion resistant basalt fiber cables with a higher tensile strength than steel might be preferred. There is plenty of aluminum on the Moon but aluminum threads strip and rip out easily. Aluminum is too soft for ball bearings and roller bearings. Steel is needed for threaded pipes, fittings, nuts, bolts and screws. Fortunately, the required mass of these will not be too large. Steel is needed in moderate quantities for threaded parts, nuts, bolts, ball bearings, roller bearings, hand tools, power tools, cutting tools (cemented carbides might also be used) drill bits (cobalt steel), machines like lathes, grinders, milling machines, extruders (for chambers and rams), gears, drive shafts, axles (titanium might suffice for that), razors and good knives. Strong maraging steel contains almost no carbon but does require cobalt (from meteoric iron-nickel fines) and molybdenum and can be used for things like rocket motor casings but it won't take a good edge for blades and knives. Nickel, chromium and manganese exist on the Moon for steel alloys like stainless steel. Steel can be customized for almost any desired application by altering carbon content, heat treating and alloying. Steel metallurgy, like chemistry, is a very mature science. Steel's only major weakness is that it rusts. That will not be a problem in the lunar vacuum. Steel can be easily recycled by melting and re-casting. Pure iron which is abundant in lunar regolith has strength properties similar to wrought iron and can be used for ornate works, nails (in aerated auto-claved concrete), hinges, pins, pots, pans, rails, handles, rods, furniture, etc. Wrought iron was used for steam train rails before the now obsolete Bessemer Process made large amounts of cheap steel available in the 19th century. In the low gravity and rust free vacuum of the Moon pure iron might be sufficient for railway tracks.

Additive manufacturing,"3D printing," with metals is commonplace today. Stainless steel, low alloy steel, maraging steel, cobalt and nickel alloys are all used presently.[21] Direct metal laser sintering can produce solid parts without using a binder and it can make parts with complex geometries that CNC milling cannot.[22] Planetary Resources and 3D Systems actually printed up a model spacecraft with powdered meteoric iron-nickel material.[23] Certainly, lunar meteoric iron-nickel particles can also be used for additive manufacturing with electron beams or lasers.

Aluminum, magnesium and titanium are also important structural metals. Silicon will be in demand for solar panels and it can also be used to alloy aluminum. Manganese and chromium are used in steel alloys. Sodium, potassium, sulfur and phosphorus are also present on the Moon. Numerous electrochemical methods for extracting these elements exist. The most efficient way to produce all these elements might be the use of Supersonic Dust Roasters and All Isotope Separators.

There is hardly any copper, zinc, lithium, molybdenum or vanadium on the Moon. The best aluminum alloys are made with copper, zinc and/or lithium. Lunar engineers will have to rely on pure aluminum and aluminum alloys of manganese, magnesium and silicon. There will be commercially pure titanium and titanium alloys. Aluminum is used in most titanium alloys. It is called an alpha stabilizer. Vanadium is also used a lot for the workhorse alloy Ti-6Al-4V. This is called an alpha-beta alloy and these are most useful. Vanadium and molybdenum are called beta stabilizers. Unfortunately, these are present only in tiny traces on the Moon. Maybe we could use a different beta stabilizer besides vanadium. The only beta stabilizers abundant on the Moon are iron, chromium, manganese or silicon.[24] It seems we might be making titanium-aluminum-iron or titanium-aluminum-silicon alloys on the Moon someday.

We can see that additional payloads to the Moon should include but not be limited to:

solar or electrical furnaces for melting and pouring basalt

small digging tool attachments for making crude sand molds in the ground

some iron starter molds for basalt

platinum-rhodium bushings and whatnot for basalt fiber drawing

heaters to sinter basalt packed into iron molds, packing tools for robots

heaters, perhaps induction heaters, to melt steel

magma electrolysis cells

metal powdering equipment (centrifugal electric arc perhaps)

rovers with low intensity magnetic separators for harvesting meteoric iron fines

centrifugal grinders

3D printers that can make heavy iron molds

carbon monoxide processing equipment

Lunar Dust Roasters and All Isotope Separators

more solar panels to power all this electrical equipment

6. Metal Massive, Unitary, Simple Things

Cold metals can be shaped by rolling, extrusion, forging and spinning. This work hardens the metal. When this is not wanted, metals can be hot worked. It takes energy to heat the metal but not as much as melting does and less horsepower is needed to roll, forge or extrude the softer hot metal. Items made of metal on the Moon include but are not limited to:

Foils: food wrapping, parasols, solar shields, reflectors to increase solar panel output, shields for protecting cryogenic tanks and other pieces of equipment from intense solar radiation, mass driver covers for thermal and dust protection

Metal Sheets: buckets, bins, tool holders, shelves, drawers, computer and electronics casings, tableware, making radio and radar dishes, trough reflectors for solar thermal power systems and dish reflectors for solar furnaces. Sheet metal can also be used for more mundane items like toolboxes, storage trays, tool racks, simple boxes, shelves, drawer cases and drawers, air ducts, cans, pots, pans and storage sheds. Sheet metal tubing could be used for heat dissipating space radiators.

Flat Plates: slusher buckets or scrapers, excavator buckets, bulldozer and road grader blades, ground vehicle parts, spacecraft frames, metal floors, fluidized beds (with some tubes and other parts), appliance parts and casings, even pots and pans by stamping small circular thin metal plates. Metal plate could also be used for vehicle frames, sand mold boxes, input hoppers, conduits, and many other things.

Curved Plates and Spun Domes: rocket propellant tanks, fuel cell reactant tanks, water tanks, oxygen (and other gases) tanks, pressurized ground and space vehicle cabins, parabolic solar trough reflectors, spun domes for radio and solar concentrator dishes. Curved metal plate could be used to make high pressure gas cylinders and skins for the pressurized habitat modules. Thicker sheets and thinner plates could be used for rocket propellant tanks

Rails, Bars, Beams: ground vehicle and earth moving equipment frames as well as other parts, building support structures, railroad tracks.

Rods: axels, "tent" or canopy poles, radio antennas, rebar, "telephone" poles for power lines and phone lines.

Wires: power lines, phone lines, electrical wiring, motor coils, steel cables for earth moving equipment.

It can be plainly seen that the simple objects made by rolling, extrusion, spinning and drawing have many uses. The heavy equipment needed for this can mass produce these items from lunar metals. Aluminum, pure iron, meteoric iron-nickel and steel will be most useful. Magnesium is soft but not very ductile unless it's hot. Titanium is hard to cold work but can be hot worked. Many titanium parts could be made from powder by electron beam fusing or sintering, a kind of 3D printing, outside in the vacuum.

Metal plates both flat and curved and metal pipes will be needed to build pressurized cabins for ground vehicles, railroad cars and spacecraft, habitat modules, rocket propellant tanks, liquid and gas storage tanks, and numerous other products.

This would call for payloads of:

    rolling mills for making metal plates, rails, beams, pipes, etc,

    centrifugal casting machine to make basalt pipes

    extruders to make metal tubes, rods, bars, etc.

However, rolling mills and extruders can be very massive pieces of equipment. A rolling mill could weigh 40,000 kilograms or more and an extruder up to 10,000 kilograms or more, although there are smaller lighter versions even bench top scale rolling mills and extruders. A Falcon Heavy rocket could put 54,400 kilograms in LEO, so the problem is not getting it up there. If 1000 metric tons of cargo was lifted to the Moon a 40 metric ton rolling mill would only be about four percent of the total. However, more than a dozen rolling mills might be needed to roll plates and sheets of various thicknesses as well as I-beams, pipes and rails. It would seem that the better way of doing things is to make heavy equipment on the Moon. This boils down to extracting iron from regolith with SDR-AIS devices, converting it to steel, sand casting large heavy parts and machining them to exact tolerances with imported CNC machines. Robots and human workers together can assemble the heavy rolling mills, extruders, forging presses, etc. Small parts can be made with 3D printers, powder metallurgy and even machining by hand. Hard steel bearings and gears come to mind. Complete machine shops with lathes, drill presses and more will be needed.

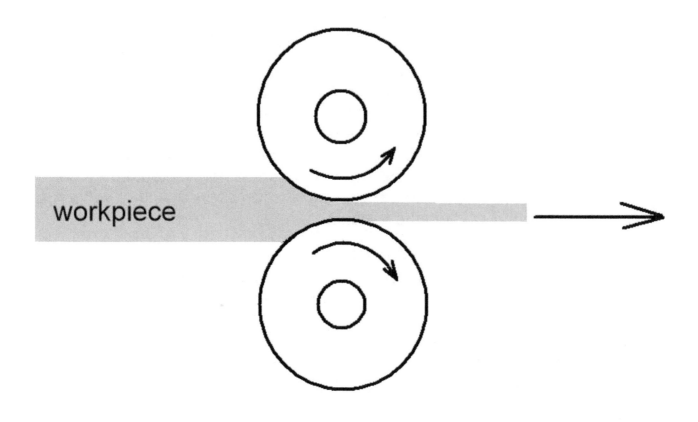

Fig. 12 Rolling flat plate or sheet

In principle, metal rolling is very simple. Thick pieces of metal are rolled between cylindrical steel rolls to make thinner pieces. If metals are rolled between a series of rolling mills in tandem, thin plates and sheets can be made. Rolls with more complicated cross sections can make I and H beams, U beams, T beams, channel beams, corner sections, angle irons and railroad rails. Red hot metals are softer and easier to roll. This also avoids strain hardening of the metal. In the vacuum of the Moon and outer space, the formation of oxide scale on the hot metals will not happen, so cleaning off the scale with strong acids (pickling) will not be necessary. Flat pieces thicker than 0.25 inches (6 mm) qualify as plates. Thinner pieces are sheets and the thinnest pieces are foils. Plates and sheets can be cut into strips, disks and complex shapes on laser cutting tables for various uses. Foils can be cut with razor sharp blades. Metal plates can also be curved with special rolling mills. Aluminum alloys will probably be used the most for flat and curved plates, sheets and foils. Fortunately, lunar highland regolith is rich in aluminum. It will be used for ground vehicles and spacecraft.

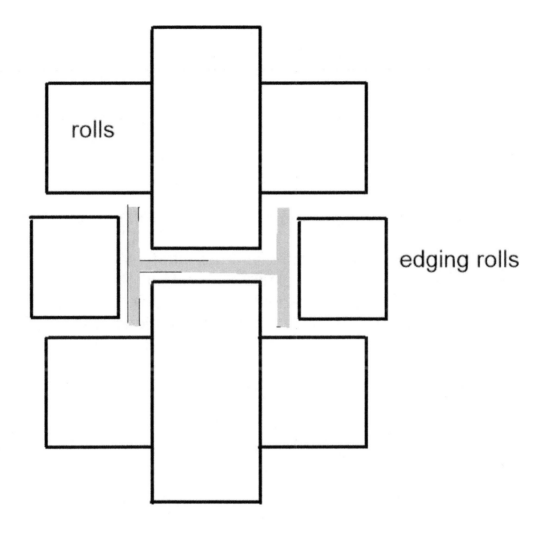

Fig. 13 Rolling I and H beams

7. Casting on the Moon

There can be no doubt that materials production devices from meteoric iron harvesting rovers to solar and electrical furnaces and other things including solar panels for power must be sent to the Moon ready to work and supply materials or nothing can happen at all. Some production devices like 3D printers must be sent up ready to work also.

Working with metals will require furnaces for melting and heat treating and heavy machines like rolling mills to work the metals into useful things. Meteoric iron containing 5% to 10% nickel and pure iron from the regolith could be converted to steel by roasting it with carbon. This is the ancient crucible steel process. The Moon doesn't have a lot of carbon but a tiny amount of carbon can make a large quantity of steel. Casting that steel into large rollers a meter or more (40 inches) in diameter and two meters (80 inches) long that are later ground and polished with CNC (computerized numerical control) machines to within two ten thousandths of an inch presents a problem. Fairly pure silica or olivine sand will be needed for expendable molds along with clay and water to bind the mold. While sand might come from regolith, clay is non-existent on the Moon and water is precious. In the vacuum the water will sublime and the wet mold will dry up. Unless this job is done inside a pressurized structure so that the water can be recovered from the air within that water will be lost.

What about using something besides wet sand and binders other than clay? Sand molds can be made with resin but the Moon doesn't have large quantities of light elements needed to make resin. Eventually there will be enough polar ice mining and solar wind implanted volatiles harvesting to get the elements needed to make resins, but at first it would be imported and recycled. The resin bonded sand mold would have to be contained in a sealed metal box so that when hot metal was poured in and the resin volatilized it could be recaptured. Sodium silicate is another potential binder. This compound can be made on the Moon but it must be dissolved in water then mixed with sand. In the vacuum the water would evaporate, so the job must be done inside a pressurized structure and the sodium silicate must be allowed to dry out before moving the mold outside and pouring metal. Making molds or cores from plaster or cement will also require a pressurized structure where water vapor is condensed from the air. There is no way to mix plaster or cement with water out in the vacuum. The water will evaporate in a flash. At least the actual metal casting can be done outside. Small parts might be cast inside, but casting large parts inside some kind of habitat module would release a lot of heat and that would demand a powerful air conditioning system. It might get so hot inside that only robots can work within.

It seems doubtful that metal objects with a mass of several tons could be cast inside large inflatable structures even with a concrete floor. Such a structure will still be needed for making sand molds bound with sodium silicate solution and for making molds and cores out of plaster or cement. Airlocks and wheeled carts or pallet jacks will be needed to get large molds outside and to get solidified castings inside.

8. Grinding Metals

Rough castings will have to be finished by grinding and polishing with CNC milling machines. It is not likely that large heavy parts like steel rollers and extruder barrels would be made entirely by CNC machining of big blocks of steel. This would wear out expensive imported tungsten carbide cutting tools. It should be possible to make castings and finish them to exacting tolerances with CNC machines.

There are CNC milling machines that can machine 60,000 kilogram (132,000 pound) workpieces, but they are as big as a two car garage.[25] On the Moon, workpieces to be shaped into rollers would only have a mass of about 9,000 to 14,000 kg. (20,000 to 30,000 pounds). In low gravity they would weigh one sixth as much and be easier to move around. There are CNC machines that could work pieces this large and they have a mass of about 16,400 kg. (36,000 pounds).[26] It would seem to make more sense for Moon miners to send up a 16 ton CNC machine that can be used to make numerous machines from rough castings rather than send up complete 40,000 kg. rolling mills and other heavy machines that cannot replicate themselves.

Machining requires lubricant. Sometimes water is used. This won't work in the vacuum. It seems CNC machining will have to be done inside the same pressurized structures where molds and cores are made. Tools will get hot but there will not be nearly as much heat released by machining as there would by melting and casting parts weighing several tons. Lubricant/coolant will be recaptured, filtered, cooled and reused over and over again. Working inside a pressurized structure will also protect CNC machines from extremes of lunar temperature. It might be possible to adapt conventional "off-the-shelf" machines designed for work on Earth to the low lunar gravity and that could save money.

9. MUS/cle

This is a situation where Peter Kokh's MUS/cle strategy can be applied.[27] "MUS" stands for massive, unitary and simple while "cle" stands for complex, lightweight and electronic. If we can just make the large heavy parts of machines on the Moon and import the electronic controls and perhaps the electric motors the cost of sending cargo to the Moon will be reduced and the industrialization of the Moon becomes more practical.

If CNC machines can make large heavy or massive, unitary and simple parts on the Moon from rough metal castings it will reduce the requirement for transporting heavy objects by rocket and amount to huge savings. The 3D printers could make small intricate steel parts needed in limited numbers including gears, bearings, molds and dies. The rolling mills and extruders would mass produce large steel and aluminum plates and sheets, beams, rails, pipes, etc. A combination of pressurized structures where molds and cores can be made along with CNC milling machines, 3D printers, machine shops with drill presses, lathes and other machine tools, solar and electric furnaces and accessory equipment in addition to some other imports should make it possible to produce heavy rolling mills, extruders, large engine lathes and forging presses on the Moon. When all this equipment is working it should be possible to make anything out of metal that has to be made of metal.

10. Manufacturing

The best of terrestrial conventional manufacturing techniques will be applied on the Moon even in the age of 3D printing. Casting is important. There could be times when casting is faster and cheaper than 3D printing; however, casting will require a pressurized foundry or a sealed metal container so that liquid metals don't evaporate into the vacuum. Liquids evaporate in a vacuum. This makes thin film physical vapor deposition (PVD) with molten metals in a vacuum possible. Free vacuum will make PVD easy on the Moon but it can complicate casting. Atoms of molten metal will not reach lunar escape velocity but there could be loss of material due to boil off from molds unless they are sealed. The casting job can't be done indoors.

Small parts made of aluminum and magnesium could be cast in plaster molds inside the foundry. Plaster, calcium sulfate, would be obtained by leaching anorthite with sulfuric acid. While basalt, steel and iron might be cast outside in the vacuum without too much loss of material by evaporation, wetted sand molds would ordinarily be required to cast these metals and that won't work out in the vacuum. Sand molds require a binder, usually clay; however, clay is formed by hydrological processes and it will not be found on the Moon, but there might be clay on Mars. It might be possible to use a little chemical magic to make synthetic clay. Polymers and sodium silicate should also be considered. Sand molds bound with dried sodium silicate are used today on Earth and the Moon has the necessary ingredients of sodium and silicon dioxide to make the stuff.

Molten metals even for small castings will emit lots of heat and a powerful cooling system will be required in the foundry in addition to concrete floors and barriers that can stand up to spilled liquid metal. Concrete is a mixture of gravel, sand and cement. Cement can be produced by roasting highland regolith at over 1500 C. to drive off SiO_2 and MgO and enrich CaO and Al_2O_3 components. If a setting time retarder is needed, some $CaSO_4$ can be made by leaching regolith with sulfuric acid. As discussed previously, a pressurized structure where molds and cores can be made with sand, binders, cement and/or plaster and machining can be done with lubricants/coolants that can be recycled is necessary. This structure or foundry could be made of inflatable Kevlar modules that have concrete floors poured within and are covered outside with regolith for radiation shielding. Cement will demand a lot of water and it can only be mixed and poured inside pressurized modules. As it dries and sets it will release most of its water into the air and condensers could recover water from the air. Water can come from polar ices. It can also be obtained by combining LUNOX with imported hydrogen or hydrogen from solar wind implanted volatiles mining. Since water is 8/9s oxygen this could be worthwhile.

Casting on Earth seems like a fairly straightforward manufacturing process. On the Moon it becomes rather highly involved. Fortunately, the need to cast anything really huge beyond heavy rolling mills, extruders and drop forging presses does not exist. Large metal things like plates and I-beams can all be made outside with rolling mills and extruders. Lunar workers could teleoperate robots that load billets of metals into machines that extrude beams for vehicle frames and weld them up outside with arc welders. Friction stir welding could be used with aluminum. A rapidly rotating ceramic tool is guided along the joint where two metal pieces meet. Friction with the spinning tool generates heat that fuses the aluminum. Much can be made with flat and curved metal plates produced by feeding ingots or slabs of metal into rolling mills. Those plates can be square or workers can laser cut them into various shapes including disks. Beams and rails can be rolled. Beams of various dimensions, rods, bars, rails, pipes, and metal fibers can also be produced by extrusion.

This is the "Lego set" lunar makers have to work with. Rods can make axles. Beams can make vehicle frames. Pipes or tubes can also make vehicle frames. Flat plates can make buckets and ore bins. Disks can be used for wheels and maybe presses can even stamp out wheels. It shouldn't be too hard to make ore cars, rails and buckets and cables for slushers with rolling mills, extruders, presses and a small foundry with machine shop along with 3D printers. A big engine lathe instead of a giant press could spin metal domes from circular metal plates, disks, outside. Beams and rods can make power cable towers and supports for reflector systems. If one is imaginative enough, it might be possible to extrude basalt. Take a billet of basalt, get it red hot and soft, and squeeze out beams for making things like towers and supports. Drill holes in the beams with lasers and bolt them together with steel bolts and one can come up with all sorts of structures. Solar furnaces will require lots of frame members to support reflectors and crucibles. Parabolic solar trough reflectors can be made by rolling and curving sheet metal and dish reflectors can be made by spinning. Hemispheres made by spinning could be welded together to make spherical storage tanks for water, LOX, LH_2 and other liquified gases. This work can all be done outside with machines mounted on solid basalt pads with parasols to shield everything from the blistering hot lunar Sun and trap warmth by night.

Forging metals will also be important when this can make parts faster and in larger quantity than 3D printing. Drop forges would have to be very tall and have massive weights in low lunar gravity. Compressed oxygen could drive forging hammers too. Electromagnetic systems are also possible. Since the Moon lacks oil and leakage into the vacuum is likely forging presses would probably not be hydraulic. All sorts of parts can be made from hot metal blanks. All varieties of steel dies might be made by 3D printing; however, printed parts are sometimes more porous and weaker than cast parts. Casting steel dies in the lunar foundry might be called for. The steel dies would be heated and water quenched to harden and temper them. Steam from the quenching of hot metals would be condensed from the air in the pressurized foundry module. Forgings will be in demand. A jet liner contains thousands of forged parts. Rockets, ground vehicles, robots, rovers, refrigeration devices, machine tools and many other things will contain forged parts.

It is true that 3D printing can make some large parts like an airplane wing, but it is slow. It wouldn't make sense to print an I-beam, which would probably be the biggest single part made of metal on the Moon besides parts for heavy equipment. An I-beam would be rolled or extruded outside. There might not be any demand for large I-beams anyway until lava tubes are sealed and pressurized and buildings are constructed within using conventional techniques. Curved plates and domes for "sausage" shaped habitat modules can be cranked out by rolling and spinning. An airlock hatch seems like something that would be cast. It could be possible to stamp or forge hatches outside if there is a big enough press and some disk shaped billets.

Lunar makers must strive to make vehicles and machines on the Moon using 3D printing, rolling, extruding, stamping, forging and only rarely casting. They will have pressurized machine shops where parts are drilled and milled with great precision. There will be assembly shops or garage modules where vehicles and machines are put together. Operating machine tools and assembling things will not generate the extensive heat that melting and casting metals will.

Fasteners will be necessary. Bolts and screws are made in a bolt rolling machine that rolls rods (made by extrusion) between two dies. Nuts and rivets are also going to be required.

To continue with the payload list for setting up a bootstrapping lunar industrial base:

    Inflatable module for foundry with powerful cooling system.

    cement mixer to make concrete for foundry floor and barriers

    electric arc and friction stir welders

3D printers as needed

more solar panels and associated hardware, wiring, etc. to power machines

very large engine lathe

cutting lasers, perhaps a cutting table

forging hammers

sulfuric acid leaching systems (these might be made of acid resistant basalt on the Moon)

machine tools (drill presses, lathes, grinders, boring and milling machines, etc.)

CNC milling machines

bolt rolling machines

spare parts for machines--could be printed up on the Moon as needed

precision instruments and hand tools

11. Construction

There will be basalt poles to support power lines and telecommunications cables. There will be basalt supports for solar panels. Solar farms will cover thousands of square meters of land. There will be melted or sintered regolith roads and railways consisting of steel rails and steel ties on beds of gravel.

Shelter is of utmost importance. It seems that it will be desirable to construct shelter on the Moon instead of continually importing inflatable Kevlar modules from Earth. Aluminum cylinders with domed ends could be used. It might even be possible to make box shaped habitat by welding up flat metal plates and welding in internal supports like webs.

Contour crafting is interesting. This is basically 3D printing on a large scale using cement. However, hydraulic cement won't work outside in the vacuum. The water will evaporate before the concrete can set. This kind of cement could be used with bricks, cement board and slabs to make steps, walls, even furnishings inside of pressurized modules. In sealed and pressurized lava tubes entire buildings could be constructed using conventional techniques. Water would be needed to make all this cement and that would come from polar ices, solar wind implanted volatiles mining for hydrogen and/or imported hydrogen combined with LUNOX as mentioned above. Sulfur cement might be used inside pressurized spaces. Outside on the Moon it might also be possible to use sulfur cement. There are substantial traces of sulfur in the regolith that can be extracted by roasting regolith at up to 1200 C. The sulfur would simply be mixed with sand (sieved regolith) and gravel, heated to 140 C. to melt the sulfur and then poured. A contour crafting gantry could be used to "print up" all sorts of structures. The only drawback is that the extreme heat of lunar day (127 C. at the equator) could melt the sulfur (m.p. 115 C.) and the construction would collapse. Bulk regolith is an excellent thermal insulator. The temperature at a depth of one meter at the equator is a constant -20 C.[28] If the work of pouring the molten sulfur cement is done behind or beneath foil shields and the solidified structure is then covered with several meters of insulating loose regolith then it should be safe.

Another possibility is the construction of fused rock or basalt structures. The contour crafting gantry would need a bucket consisting of a high temperature resistant metal like molybdenum or tungsten in which regolith was melted and oozed out to form domes of rock layer by layer with layers a few centimeters thick. It might also be possible to use a nickel-steel bucket with an active cooling system. Fused rock structures should be very strong and the material needed to make them is just sitting on the Moon waiting to be excavated. A contour crafting gantry seems like a heavy cargo to the Moon. It seems likely that Moon miners will have the steel beams and other parts needed to build contour crafting gantries plural for doing all kinds of work on the Moon. Two more additions to the lunar industrial seed would be:

> Contour crafting gantries, or just the components that cannot be made on the Moon

> Crucibles made of molybdenum or tungsten with heating elements

Fused basalt would melt at 1150 to 1350 C. This material will have no trouble standing up to the heat of the lunar day. However, thermal cycling between night and day could lead to cracking. Layers of loose regolith several meters thick would cover the structures to provide cosmic ray shielding and more than enough thermal insulation that would prevent thermal cycling. Basalt doesn't have a lot of tensile strength compared to metals so module walls might be as much as a foot thick as opposed to mere fractions of an inch that would be needed with metals. Fused basalt or highland regolith would be easy to excavate, screen, and load into contour crafting gantries. No metal extraction, metal working and welding would be involved. This could be a much cheaper way to make habitat modules and it is easily automated. Energy from solar panels will be needed to melt the material but that shouldn't be a showstopper.

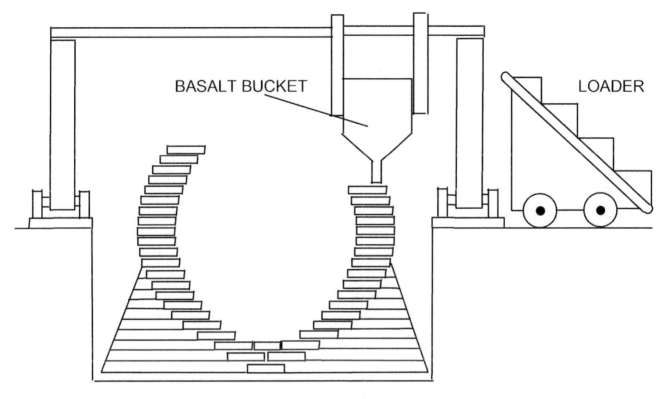

Fig. 14 Cross sectional diagram of machine printing cylindrical habitat with extruded molten basalt.

12. Motors

At first, equipment from Earth will be complete with motors. There will be electric motors galore on the Moon in a range of sizes. They will be used for vehicles, digging machines, contour crafting gantries, slushers, railways, rolling mills, extruders, forging hammers, compressors, refrigeration equipment, ventilation fans, water pumps, sewage pumps, 3D printers, small electronics, spacesuit backpack mechanisms, solar panel and antenna tracking systems, coolant pumps, appliances, sewing machines and looms, power tools, machine tools, farm equipment, even electric razors and toothbrushes, etc. A substantial part of lunar industry will involve electric motor production. Aluminum wire will probably be used since copper is so rare on the Moon. Sparse carbon and hydrogen will be combined with plentiful silicon and oxygen to make silicone insulation for the motor windings. Iron and steel will be available. Motors might run on DC or direct current from solar panels might be inverted to 3 phase AC to run 3 phase motors. Such motors can be lighter and more powerful. Entrepreneurs who set up a factory on the Moon that makes a wide variety of electric motors and replacement parts from lunar sourced materials will have a reliable market with a real future. Another payload for the lunar "seed:"

Electric motor winding machines and any specialized tools needed to get electric motor production going on the Moon

13. Solar Panels

Lunar industry will have a voracious appetite for electrical energy. It would seem reasonable then that solar panels and related gear should be made on the Moon. Dr. Peter Schubert has proposed growing ribbons of doped silicon between carbon blocks in the vacuum. These would be cut to size with diamond saws or lasers. He has also designed a device that works in a manner similar to a mass spectrometer. The device can produce oxygen, silicon, silicon doped with phosphorus, aluminum and iron; possibly other elements too.[29] This device does not require imported chemical reagents like chlorine and fluorine. It consists of some exotic materials that would have to be imported like thorium oxide and platinum-rhodium. Much of the machine, like the piping systems, waste heat radiators and cryo-chillers could be made on the Moon from steel and ceramic materials. Boron for p-type silicon is rare on the Moon. Aluminum could be used instead. Phosphorus for n-type material is available. Aluminum could also be used for backing and wiring. Glass could be produced for anti-reflection coatings. More lunar industrial seed cargoes could include:

Complete and partial Supersonic Dust Roasters and All Isotope Separators

Carbon blocks, diamond saws, lasers

If for some reason this device fails to deliver, then silicon might be produced by serial magma electrolysis. This could be zone refined to a high degree of purity. If this fails, then silicon extraction will require the use of corrosive fluorine imported from Earth. Fluorine could be shipped easily in salt form and electrolyzed to free up the halogen.[30] This will increase costs; however, the cost of imported fluorine will not be as great as the cost of imported solar panels given the huge need for them.

14. Railroads

Dirt roads on the Moon will mean slow going. Cast or sintered basalt roads would probably be too slick to be practical for high speed travel. Basalt roads would probably buckle and crack due to the temperature extremes of the lunar day and night. Faster travel for large numbers of people or heavy loads of raw materials and finished products could be achieved with railroads. When industry has grown large enough to produce thousands of miles of railroad tracks, solar power plants, electric locomotives and various kinds of cars, railway systems will be built on the Moon to link settlements and mine sites together.

The first railroads on Earth were built before the Bessemer Process made large quantities of cheap steel available. Wrought iron was used for tracks. Pure iron can be produced on the Moon in large amounts and it has similar properties to wrought iron. Both have about 40,000 psi compressive strength and tensile strength. Perhaps in the low gravity of the Moon pure iron would be sufficient for railroad tracks. If steel is required carbon will be needed, but a small amount of carbon can make a lot of steel. With 80 tons of carbon from volatiles harvesting it will be possible to make 24,000 tons of 0.33% carbon mild steel. European rails are 40 kg. to 60 kg. per meter, so this could make 240 kilometers of track.[31] There will be no rust on the Moon but high temperatures by day will mean that the tracks will have to have gaps to allow expansion and solar shields made of foil or sheet metal.[32] At night the iron or steel could become so brittle in the super cold that the metal will crack. Electric heating elements will be needed to keep the rails warm at night. The trains will be powered by a third rail or an overhead wire that could electrify rail heaters as well as locomotives.

Railroad cars could be made of curved aluminum alloy plates and domes. Air wants to fill out a spherical volume. The sausage shaped cars will be stronger to contain air pressure. Similar cars could contain liquid freight like water from the polar regions. Flat metal plates could be used to make flat bed cars and unpressurized box cars for freight. Wheels could be made of cast alloy steel and under carriages could be made of lightweight titanium. Electric motors would use aluminum wire for coils.

Rail trips across the Moon will probably only last a matter of hours or days. Exposure to cosmic rays will be brief. There's no way to make a railroad car surrounded by 9 tons of regolith per square meter of hull. For people not concerned with sight seeing there could be radiation safe subways. Surface railroad trains will still need solar flare shelters. Fortunately, shielding for solar flares does not need to be nearly as massive as shielding for cosmic rays. A layer of water or polyethylene about 12 cm thick can reduce solar flare radiation doses to safe levels. Some train cars will have water and sewage tanks in the walls along with fresh food supplies and some plastic shielding. Since food is mostly water it can serve as a good radiation shield. If there is a solar flare, passengers can crowd into the shielded supply cars for a few hours.

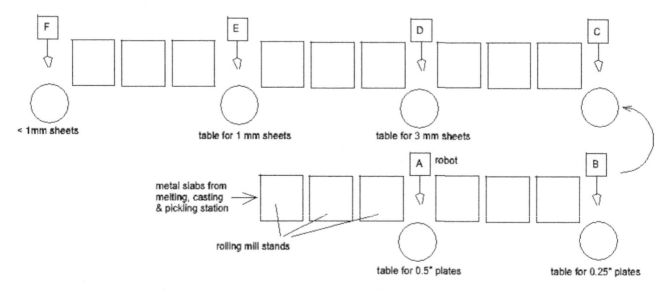

Fig. 15 Rolling mill stands in series with robot attendants could make metal sheets and plates of varying thicknesses in large quantities. These would be used to make railroad cars and solar shields for tracks. Mills could also make massive numbers of iron or steel rails for tracks.

15. Replication

It will not be cost effective to keep importing heavy production machines, robots, bulldozers, habitat, etc. These things must be made on the Moon in greater and greater numbers as lunar industry and populations grow. The Moon cannot "cut the cord" with Mother Earth completely. Factories with clean rooms where integrated circuit chips are made cost billions of dollars. Computer chips will come from Earth. In the more distant future such factories might emerge on the Moon. Until then the mechanical parts for robots, vehicles and other machines will be made on the Moon while the electronic brains are imported.

The various manufacturing methods like 3D printing, rolling, extruding, forging, machining by hand and with CNC (computerized numerical controlled) machines and casting will be used. To replicate some machines large part casting might be required. An inflatable made of Kevlar with more tensile strength than steel with a concrete floor could do small casting jobs, but large parts will have to be cast outside in polymer or sodium silicate bound sand molds. Some carbon and water would not be sacrificed if the mold is sealed. Loss of water or organic chemicals can be avoided if lava tubes of reasonable size can be found, sealed and pressurized with oxygen. Within a lava tube it would be safe to pour hundreds of pounds of molten metals and recover water vapor from the sand molds with dehumidifiers. Living quarters, offices and workshops could be bricked up inside of lava tubes to house hundreds, perhaps thousands, of humans someday.

16. Hypothetical Materials

Ferrosilicon from magma electrolysis might serve as a low performance rocket fuel after powdering. Iron and silicon burn furiously in pure oxygen. Monopropellant might be made from a mixture of ferrosilicon and liquid oxygen. Bipropellant rockets using LOX and FeSi mixed with a carrier liquid, perhaps silane, might be more reliable and safer. Ferrosilicon might not be as powerful a fuel as aluminum but it will be much easier to produce. It can be acquired along with oxygen by simple magma electrolysis. Monopropellants consisting of aluminum powder and LOX have been studied.[33] Ferrosilicon deserves more investigation.

Basalt fibers bound with polymer resins are now being used to reinforce concrete. Resins will not be plentiful on the Moon. There will be some carbon for steel and some for agriculture and some for organic chemicals, but those organics will be pricy. Glass fiber reinforced glass matrix composites have been suggested. Unfortunately, very little work has been done with this material. The pure silica fibers would add tensile strength and fracture resistance to a matrix of glass that has been doped with sodium and calcium to lower its melting point.[34] Why not a basalt fiber reinforced basalt matrix composite? The matrix could have its melting point reduced so as not to melt the fibers by doping with sodium, potassium, calcium and/or magnesium. It seems that this material would be "easier" to produce than glass fiber reinforced glass composites. Basalt is readily mined and melted. Moon miners have to look at lunar basalt compositions to. Lunar basalt has more iron in it than terrestrial basalt. It might be desirable to alter iron contents with magnetic separations. Casting basalt and fiber drawing in the vacuum and low gravity must be studied. On Earth, makers of basalt products cool the fibers with a water spray. In space, a sealed chamber and a spray of cooled helium gas that is recycled might be called for to cool the fibers.

If basalt fiber reinforced basalt matrix composites are feasible and cheap, then this could be a material for space frames for power satellites, space stations, space shipyards, large telecommunications platforms and space telescopes. No imported reagents would be needed. Manufacturing this stuff, which would have a density of about 2.95 vs 2.7 for aluminum, in quantities needed for solar power satellite construction should be cheaper then producing aluminum.

The primary structure of a powersat is about ten percent of its total mass or about 5,000 tons for a 50,000 ton 5GWe SPS and 10,000 tons for a 100,000 ton 10 GWe satellite.[35] If 20 powersats a year are built and each is rated at 10 GWe then in 50 years there would be 1000 of them and 10 TW of power. That's a vast amount of energy. About 200,000 tons of aluminum would be needed annually just for SPS frames. This would require quite a bit of infrastructure on the Moon. Fiberglass or basalt might be cheaper. If basalt fiber reinforced basalt composites are possible and cheap, this would drastically increase the value of lunar basalt and make the case for a mare/highland coast installation much stronger. Basalt composites reinforced with silica fibers or fibers made from melted and drawn anorthositic highland regolith should also be investigated.

Insignificant traces of copper exist in regolith. Calcium conductors might serve as an alternative to aluminum. Calcium is a metal not used on Earth because it will react with oxygen and moisture in the air and it can even spontaneously ignite. In the lunar vacuum this will not be a problem. Based on density, calcium is a better conductor than copper or aluminum.[36]

Lightweight calcium cables might be used for a lunar power grid with solar power plants at least 90 degrees east and west of the main settlement for a constant power supply. Direct current from solar panels will be inverted to AC, stepped up with transformers and delivered to distant locations. Low gravity and low mass calcium cables could make long distance transmission through thick cables with cast basalt utility poles more practical.

17. Mass Driver Lunar Launchers

Mass drivers on the Moon will use superconducting coils. Superconducting wire consisting of yttrium, barium, copper and oxygen or bismuth, strontium, calcium, copper and oxygen cooled by liquid nitrogen will be imported since these elements are so rare on the Moon. The coil supports could be made on the Moon from non-magnetic aluminum, titanium or cast basalt. Solar panels to power the mass driver could also be made on the Moon from available silicon, aluminum, glass and phosphorus. Power electronics will probably also be imported. Making the mass drivers partly out of lunar materials should save a significant amount of payload mass for the industrial seed.

Loose regolith could be launched in bags made of woven basalt fibers. It might also be possible to make spheres of sintered regolith. That might be simpler than making all those bags. Machinery will be needed to load the mass driver with a 40 kg. payload every second. This machinery might also be made on the Moon with only the necessary electronics imported.

If the mass drivers can be kept warm enough to prevent cracking of the metal parts at night and if power can be obtained they could operate constantly. At least two mass drivers would be desirable so that one could shut down for maintenance while the other keeps working. If a mass driver breaks down the other one can keep the payloads going. Shutting down at night would be very inefficient. The only ways to work 24/7 are a small nuclear powerplant, vast fields of solar panels and power storage systems, distant solar powerplants about 120 degrees East and West of the base and long distance cables or a powersat at EML1. Nuclear power is probably out for political reasons. A powersat at L1 seems unlikely without space infrastructure there. This leaves us with power storage systems that can store up enough energy for two weeks at a time and distant power stations. It might be possible to deploy a reflector satellite at L1 and beam power from distant stations to the base. This is a problem that will require careful analysis.

18. The Bottom Line

It could be true that history is governed by economic factors. If so, then lunar development will be governed by costs and profit margins, especially if the Moon is developed by private entrepreneurs who have to keep an eye on the bottom line. It seems reasonable that products that can be produced on the Moon with lunar available resources alone will be cheaper than products that rely on imports, but reality and the marketplace can defy common sense. For instance, it seems that it would be cheaper to pump up oil from deep wells in the USA rather than ship oil in supertankers across thousands of miles of ocean, but it's not. Presently, there is no way to predict any of the costs of doing business in outer space. Rocket launches cost tens of millions of dollars just to reach low Earth orbit and they explode frequently. Reliability must be improved, especially if large numbers of people are going to travel by rocket into space. Prices for rocket launches to LEO will probably come down in the future. That seems to be the trend for so many products be they aluminum, computers or automobiles and microwave ovens. Even if the price for a rocket launch comes down by a factor of ten to one hundred, it will still be expensive to travel in space and the use of on-site materials and energy, ISRU, will still be preferred.

Chapter 5: Special Lunar Materials and Products

Numerous lunar materials will find very interesting uses. For instance:

Magnesium

There is about as much magnesium in lunar regolith as there is aluminum. It would seem foolish not to use it. This lightweight relatively strong metal will find many uses in the low gravity of the Moon while posing no combustion dangers in the vacuum. Since magnesium can be produced with just lunar available reagents like FeSi from magma electrolysis it might be cheaper than other metals like aluminum and titanium. It can be used for wheels, vehicle frames and structural applications. It is slightly more reflective than aluminum and sheets or foils of magnesium can be used for solar collectors and solar shields. Magnesium powder mixed with LUNOX to form a slurry is shock and vibration sensitive and will detonate so it won't make a good rocket monopropellant but it could be used as an explosive. Slurries of Mg and LUNOX could be mixed up in magnesium or perhaps basalt tanks and be set off with an electric spark for blasting into solid rock. There isn't enough nitrogen on the Moon to spare for nitrate explosives. This element will be reserved for life support. We can still blast with magnesium based explosives.

Native Glass

There are glass particles in the regolith that can be electrostatically extracted and melted down to make glass products; perhaps even foamed glass for structures. There are fields of pyroclastic (volcanic) glass on the Moon. The glass particles could be extracted while mining massive tonnages of regolith in these deposits. The pyroclastic glass particles or beads if you will are coated with traces of iron, nickel, sulfur, copper, zinc, gallium and chlorine.[37] We could heat the glass beads and evaporate these lunar rare elements from their surfaces or wash off the elements with a liquid of some sort. Glass from regolith and glass from the processing of regolith for metals could be used to make glass fiber reinforced glass matrix composites. The melting point of the glass matrix would be reduced by adding Na, K, MgO and $CaAl_2O_4$.

Basalt

This is one of the most abundant and easily obtained resources on the Moon. The mare consist of pulverized basalt. All we need to do is mine up mare "soil" and press it into iron molds and sinter it to make bricks, blocks, tiles, slabs, etc. We can also melt it at 1150 to 1350 C and pour it into iron molds to cast solid items. Basalt pipes can be made by centrifugal casting. Basalt is very hard (harder than steel) and abrasion resistant. It has high compression strength. Basalt fibers could be drawn through platinum-rhodium bushings, like glass, to make basalt fiber reinforced basalt matrix composites. The basalt matrix m.p. would be reduced by adding Na and K. This material is only hypothetical. If it is possible to make this stuff it will be cheap and abundant and it could substitute for metals in many structural applications. Basalt is a bit denser than aluminum and basalt fiber/basalt matrix composites might be used instead of aluminum for space solar power satellite frames. Basalt has also been used to make brake pads. Since asbestos will not be found on the Moon, vehicle brakes could be made with basalt instead. Basalt can be carved into beautiful items. We'll need some imported tungsten carbide chisels to do that.

Calcium

There's plenty of calcium on the Moon. On the average, it composes about 8% of the regolith. This soft lightweight metal ignites spontaneously in air. This will not be a problem out-vac on the Moon, although it could sublimate in the vacuum so cladding with aluminum will be desirable. Calcium is a better conductor than aluminum and copper. It could be used for long distance power lines. Much research on calcium metallurgy needs to be done.

Calcium compounds like CaO, $CaAl_2O_4$ and $CaSiO_3$ could be used for cement and mortar for indoor applications. Calcium sulfate, $CaSO_4$, when wetted is plaster that can be used for metal casting molds, medical and dental casts, frescoes and reliefs. When wetted and placed between two sheets of woven glass fiber or basalt fiber fabric it can make drywall or sheetrock for walls. Calcium sulfate is also used as a cement setting time retardant. It can be produced by leaching anorthositic regolith with sulfuric acid.

Meteoric Iron Fines

Meteoric particles consisting of iron, 90-95%, nickel, 5-10%, cobalt, 0.2%, and lesser amounts of germanium, gallium and platinum group metals (PGMs) are present all over the Moon in a few tenths of a percent of the regolith. These can be harvested magnetically by machines that process millions of tons of regolith every year. Since the particles are fused with silicates they must be ground and processed magnetically to get pure metal. The iron and nickel can be extracted with recycled carbon monoxide gas. Separation with a device that resembles a mass spectrometer is also possible. Nickel is especially useful. It can be used to make carbonless maraging steel for rocket motor casings. Nickel can also be used to make steel alloys that are strong yet ductile and have high corrosion resistance. This will be very important for equipment that must handle hot oxygen or steam and work at extreme temperatures. It can be used as a catalyst in Sabatier reactors to combine hydrogen, carbon monoxide and carbon dioxide from solar wind implanted volatiles harvesting and/or polar ices of cometary origin to make methane. Methane can be decomposed at 900 C. to get pure carbon and recover hydrogen. Nickel can also be used as an electrode material in iron-nickel alkaline batteries like those invented by Edison. These batteries are very rugged and last for decades. Cobalt in combination with chromium, molybdenum, vanadium and/or tungsten is used to make high speed steels for drill bits and cutting tools. These will be needed in machine shops. Chromium could eventually be extracted from regolith since it contains the mineral chromite while small quantities of Mo, V and W can be imported.

Purification

Metals produced on the Moon will probably contain impurities. In the free vacuum and low gravity of the Moon purification will be easier. There will be no air therefore there will be no contamination by oxygen, nitrogen or moisture when red hot metals are purified. Rods or bars of iron, titanium, aluminum, magnesium or calcium can be zone refined to high degrees of purity in the low gravity without danger of the rods or bars falling apart. In the vacuum, the molten zone in metals being zone refined will not react with air. Impurities that have lower melting/boiling points than the metal they are dissolved in could simply be boiled off in the vacuum. In the case of iron, impurities of Na, K, S and P have much lower boiling points than the melting point of iron.

Chapter 6: Lunar Chemistry

Numerous substances like hydrogen, carbon monoxide and nitrogen can be obtained by solar wind implanted volatiles mining and polar ice mining. These can be reacted to make various chemicals and plastics.

$2H_2 + CO ==> CH_3OH$ $ZnO\text{-}Cr_2O_3$ catalyst used
                methanol

$CH_3OH + CO => CH_3COOH$ 50 atm rhodium catalyst 200° C.
                acetic acid

$CH_3OH + HCl ==> CH_3Cl + H_2O$
                methyl chloride

$3H_2 + CO ==> CH_4 + H_2O$ Ni catalyst used
                methane

$4H_2 + 2CO ==> C_2H_4 + 2H_2O$
                ethylene

$5H_2 + 2CO ==> C_2H_6 + 2H_2O$
                ethane

$6H_2 + 3\ CO ==> C_3H_6 + 3H_2O$
                propylene

So what does it all mean?

Methanol can be used as a solvent and it can be converted to formaldehyde which is used to make resins and plastics. In the commonly used formox process, methanol and oxygen react at ca. 250–400 °C in the presence of iron oxide in combination with molybdenum and/or vanadium.[38] This produces formaldehyde according to the chemical equation:

$2\ CH_3OH + O_2 \rightarrow 2\ CH_2O + 2\ H_2O$

                formaldehyde

Methanol can be reacted with HCl to make methyl chloride which can be reacted with silicon at 300° C. in the presence of a copper catalyst to make dimethyldichlorosilane $(CH_3)_2SiCl_2$. This can then be reacted with water to make silicone polymers and HCl which can be recycled because chlorine is rare on the Moon. Silicones can make lubricant oils, greases, caulk, waxes, rubber.[39]

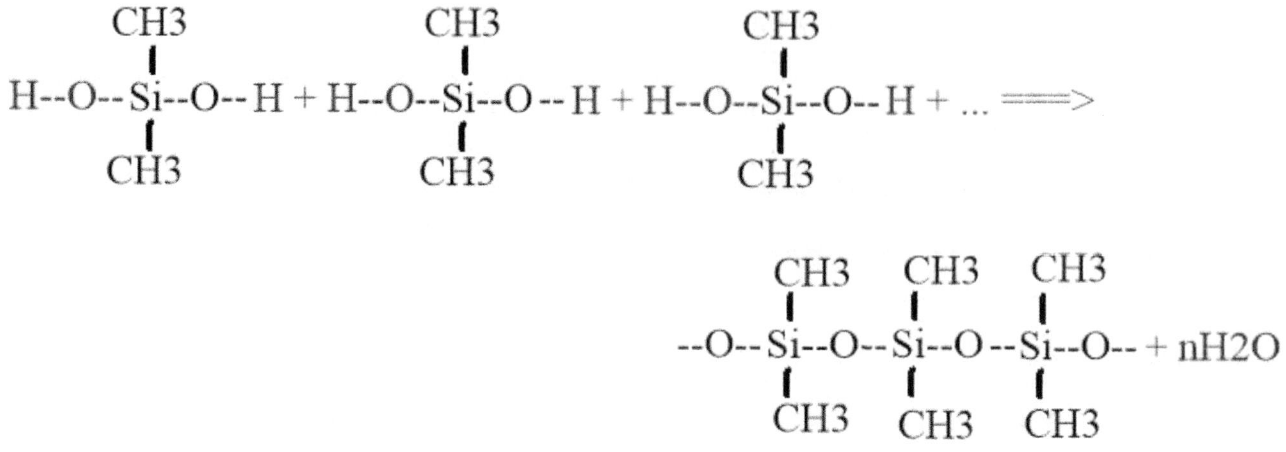

$$\text{H--O--}\underset{\underset{\displaystyle CH_3}{|}}{\overset{\overset{\displaystyle CH_3}{|}}{Si}}\text{--O--H} + \text{H--O--}\underset{\underset{\displaystyle CH_3}{|}}{\overset{\overset{\displaystyle CH_3}{|}}{Si}}\text{--O--H} + \text{H--O--}\underset{\underset{\displaystyle CH_3}{|}}{\overset{\overset{\displaystyle CH_3}{|}}{Si}}\text{--O--H} + ... \Longrightarrow$$

$$\text{--O--}\underset{\underset{\displaystyle CH_3}{|}}{\overset{\overset{\displaystyle CH_3}{|}}{Si}}\text{--O--}\underset{\underset{\displaystyle CH_3}{|}}{\overset{\overset{\displaystyle CH_3}{|}}{Si}}\text{--O--}\underset{\underset{\displaystyle CH_3}{|}}{\overset{\overset{\displaystyle CH_3}{|}}{Si}}\text{--O--} + nH_2O$$

Silicone polymer

Fig. 16 $(CH_3)_2SiCl_2$ reacts with water to form $(CH_3)_2Si(OH)_2$ which then condenses to form silicone polymers.

Methane is the most common component of natural gas. It's mostly useful as a fuel that has a much higher boiling point than LH_2. It could be reacted with oxygen in fuel cells to make electricity for motors and waste heat that might be useful for warming pressurized cabins. The products are water and CO_2. These would have to be recaptured and recycled on the Moon because they aren't that plentiful even if we do mine polar ices. On Mars where CO_2 can be obtained by pumping down atmosphere this might not be as critical. Since CO and CO_2 are very stable and decompose at very high temperatures, it is easier to combine them with hydrogen to make methane which can then be decomposed at 900 C. to get pure carbon and hydrogen that can be recycled. Pure carbon can be used to make steel and be used for air and water filters.

Ethane is the second most common component of natural gas. Like methane, it won't be found underground on the Moon but there might be some in polar ices since it has been detected in comets. It's mostly used to make ethylene by steam cracking.[40] On the Moon ethylene could be made by directly combining hydrogen and CO in the right proportions without making ethane first. Ethane can be used as a refrigerant in cryogenic systems and it can be used in heat pipes that run too cold for ammonia.[41]

Ethylene and propylene are used to make polymers. Polyethylene and polypropylene are the two most common plastics in use today. Ethylene and propylene gases are compressed to high pressure in the presence of catalysts and polymerize. Plastics would not be discarded on the Moon or in space. They would be rigorously reused and recycled. Carbon, hydrogen, nitrogen and organic compounds from polar ice will be not be cheap enough for disposables; and why trash the Moon as we have trashed the oceans?

Methanol can also be converted to acetic acid which is used to make acetate salts like calcium acetate.[42] Calcium acetate is made by reacting calcium metal, calcium hydroxide or calcium oxide with concentrated acetic acid.

Calcium acetate can be dry distilled to make acetone, another solvent that also makes nail polish remover, and calcium carbonate a.k.a. limestone. Limestone is not present on the Moon so it would have to be synthesized. It is useful as a flux in steel making and it can be used as a pH buffer in closed ecological life support systems. Acetone is also used for making numerous organic chemicals.

$$Ca(CH_3COO)_2 \rightarrow CaO(s) + CO_2(g) + (CH_3)_2CO$$

calcium acetate   acetone

$$(CH_3COO)_2Ca \xrightarrow{\Delta} CH_3COCH_3 + CaCO_3$$

Acetone   calcium carbonate

(limestone)

Acetone can dissolve plastics and some synthetic fibers. It is used as paint thinner and it is used in some paints and varnishes. It is a good solvent and can be used as heavy-duty degreaser. It is used as a solvent for vinyl and acrylic resins, lacquers, alkyd paints, and inks.[43] Acetone is used to make Plexiglas, polycarbonates, polyurethanes and expoxy resins.[44]

Most acetone today is made by the cumene process.[45] Benzene and propylene are reacted to form cumene which then reacts with oxygen in the air to form acetone and phenol. Phenol can be combined with formaldehyde to make Bakelite, a hard scratch resistant electrically and thermally insulating plastic that is seldom used today.[46] As described earlier, propylene can be made by combining carbon monoxide and hydrogen. Benzene is common in crude oil and coal tar on Earth but some might be found in lunar polar ices since benzene was detected in comet 67P/Churyumov-Gerasimenko by the Rosetta spacecraft along with toluene, butane, pentane, hexane and heptane.[47]

If benzene cannot be produced from ice it can be prepared from ethyne by the process of cyclic polymerization. In this process, ethyne is passed through a red hot iron tube at 873 K. The ethyne molecule then undergoes cyclic polymerization to form benzene.[48] Ethyne is more commonly called acetylene. Acetylene can be made by reacting calcium carbide with water (see below).

$$H-C\equiv C-H$$

ethyne (acetylene)

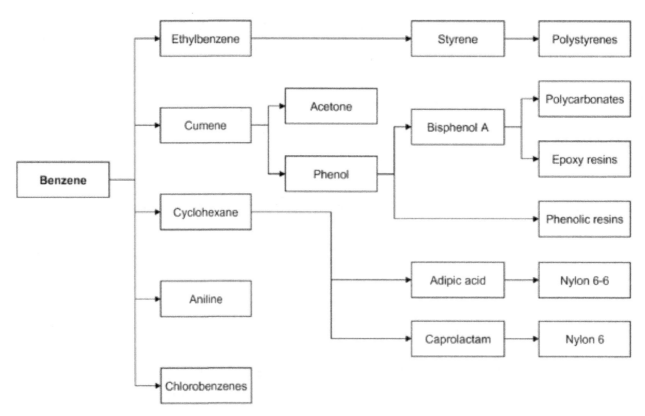

Fig. 17 Benzene has many uses.

$CaC_2 + 2H_2O \rightarrow Ca(OH)_2 + C_2H_2$
calcium carbide   acetylene

Calcium carbide reacts with water to make acetylene. Acetylene is used for Oxv-Acetylene welding and cutting. Acetylene gas is unstable and must be dissolved in acetone under pressure.

Calcium carbide is produced industrially in an electric arc furnace from a mixture of lime and coke at approximately 2,200 °C (3,990 °F).[49] This method has not changed since its invention in 1892:

$CaO + 3\,C \rightarrow CaC_2 + CO$
               calcium carbide

Calcium is present on the Moon in the mineral anorthosite $CaAl_2Si_2O_8$ which composes most of the lunar highlands regolith. Anorthosite can be melted, quenched with LOX and ground fine in rod and ball mills then leached with sulfuric acid H_2SO_4 to form a solution of highly soluble aluminum sulfate and a precipitate of silicic acid and barely soluble calcium sulfate. These can be dried out and separated electrostatically. The silica gel can be used as is, melted to make glass or roasted with sodium oxide to make sodium silicate. The calcium sulfate is plaster for walls and molds. Plaster can be applied between sheets of woven glass fiber or perhaps basalt fiber to make drywall also known as sheetrock. The $CaSO_4$ can be roasted at about 1500° C. to form solid calcium oxide and gaseous oxides of sulfur. CaO could be electrolyzed in FFC cells to get calcium metal and oxygen.[50]

Acetylene reacts with anhydrous hydrogen chloride gas over a mercuric chloride catalyst to give vinyl chloride:

$C_2H_2 + HCl \rightarrow CH_2=CHCl$
                       vinyl chloride

When heated to 500 °C at 15–30 atm (1.5 to 3 MPa) pressure, dichloroethane decomposes to produce vinyl chloride and anhydrous HCl.

$ClCH_2CH_2Cl \text{------}\rightarrow CH_2=CHCl + HCl$
dichloroethane   vinyl chloride

Due to the relatively low cost of dichloroethane compared to acetylene, most vinyl chloride has been produced via this technique since the late 1950s.[51] Nearly 20 million tons of 1,2-dichloroethane are produced in the United States, Western Europe, and Japan annually.[52] Production is primarily achieved through the iron(III) chloride-catalysed reaction of ethylene and chlorine:

$$H_2C=CH_2 \text{ (g)} + Cl_2 \text{ (g)} \rightarrow ClCH_2–CH_2Cl \text{ (l)}$$
ethylene   dichloroethane

Vinyl chloride is polymerized to make PVC or "Vinyl" plastic which was once used to make LP records and furniture including car seats. Polyvinyl chloride abbreviated PVC is the world's third-most widely produced synthetic plastic polymer, after polyethylene and polypropylene. About 40 million tonnes are produced per year.[53]

Lunar Chemistry's Future

Some might ask, why produce plastics when you have plenty of metal, glass and basalt? We need synthetic fibers, rubber washers, plastic O-ring seals, flexible hoses, sealants, caulks, gaskets, toys, space helmets (polycarbonate), space suits, electrical wire insulation, varnish for electric motor coils, rubber gloves and PPE, disposable medical supplies, food wrapping and lots of other things. These would all be reused or recycled.

The Moon is not nearly as rich in light elements like hydrogen, carbon, chlorine and nitrogen as the Earth is. Many terrestrial chemicals are derived directly from oil. Others are produced from coal, natural gas or biowaste. Nitrogen can be liquefied out of the air and chlorine can be obtained from salt deposits and seawater in abundance.

Mars might not have oil, gas and coal but it does have plenty of water in permafrost, a thin atmosphere of carbon dioxide and nitrogen, and chlorine bearing perchlorates in the regolith of the red planet. Martians could make use of some of the same chemical strategies that will be used on the Moon.

While the Moon might not ever produce enough chemicals and plastics for populations of billions, or even mere millions, there should be enough HCN from solar wind implanted volatiles and polar ices to make enough chemicals and plastics for industry in space. In the more distant future, light elements on the Moon and in space will be derived from asteroid hydrocarbons. Carbonaceous chondrite asteroids contain a tarry substance that resembles kerogen along with significant amounts of water. Rather than mine all the regolith of the Moon and extract all the polar ice it would seem wise to conserve those resources for future generations and make use of water and hydrocarbons from asteroids when space industry has grown large enough to economically mine those asteroids and transport the materials through space in interplanetary cargo and tanker ships.

Chapter 7: Biomaterials

Space farms can supply more than just food. Useful substances can be made from crops. One of the first things we can think of are vegetable oils. Many vegetable oils are used to make paints, lubricants, hydraulic fluid, soaps, skin products, candles, perfumes and other personal care and cosmetic products. Oils can be pressed out of algae, soybeans, corn, sunflower seeds, cotton seeds, hemp seeds, flax seeds (linseed oil), jojoba seeds and castor beans. Using these oils for industry will take carbon out of the CELSS loop so we will have to add some CO_2 from polar ice or solar wind implanted volatiles mining to the habitat atmosphere to maintain CO_2. levels for plants and algae.

Soap can be made by reacting oils with strong bases like sodium hydroxide (lye) or potassium hydroxide. Animal fats can also be used to make soap. The base reacts with oils and fats to form glycerin and salts of stearic, palmitic and oleic fatty acids.[54]

$$(C_{17}H_{35}COO)_3C_3H_5 + 3\ NaOH ==> 3\ C_{17}H_{35}COONa + C_3H_5(OH)_3$$
 Glyceryl Stearate   Sodium stearate   Glycerin

Soap is separated from glycerin and water by adding NaCl. This causes the soap to "salt out." It floats on top of the glycerin and water and forms a crust that is removed, dried and pressed into cakes. Glycerin can be used for lotions, cough
syrups, elixirs and expectorants, toothpaste, mouthwashes, skin care products, shaving cream, hair care products, soaps, and water-based personal lubricants. Glycerin is also used in blood banking to preserve red blood cells prior to freezing.[55]

Vegetable oils are biodegradable and have high flash points. They also oxidize easily. Castor oil is more resistant to oxidation than other vegetable oils. Castor oil and its derivatives are used in the manufacturing of soaps, lubricants, hydraulic and brake fluids, paints, dyes, coatings, inks, cold resistant plastics, waxes and
polishes, nylon, pharmaceuticals and perfumes.[56] Hydraulic and brake fluids will not be exposed to oxygen or high temperatures so the lack of oxidative stability for vegetable oils is less problematic in these applications. Jojoba oil or wax has better oxidation resistance than most vegetable oils but not as much as castor oil. In 1943, natural resources of the U.S, including jojoba oil, were used during war as additives to motor oil, transmission oil and differential gear oil. Machine guns were lubricated and maintained with jojoba.[57]

Bioplastics can also be made. Polylactic Acid (PLA) is the most common bioplastic today. It is made by fermenting corn starch or sugar to make lactic acid which is then polymerized sometimes with the help of a zeolite catalyst.[58] The main shortcoming of PLA is its low glass transition temperature. This is the temperature at which it will transform from a rigid or glass-like substance to a soft and viscous material. For PLA this happens at 111 F. to 145 F. Boiling hot drinks or a hot car in the summer could cause it to soften and deform. PLA melts at 157 C. to 170 C. or 315 F. to 338 F. PLA is biodegradable and compostable.[59] Used dirty PLA items can just be macerated and tossed in the garbage, compost heap or bioreactors.

Corn starch can be extracted by milling, grinding, washing and drying corn kernels. Starch can be mixed with water, glycerin and acetic acid then heated to form a bioplastic similar to PLA that can be poured into molds. Agar extracted from algae can be mixed with water and glycerin in a similar manner to produce bioplastic. Alternatively potato starch can be used. Potato starch is easier to make than corn starch. Simply grate some potatoes. Soak them in water. Strain off the water and let the solution dry leaving starch. [60]

Paper can be made from almost any kind of plant fiber. Rice, hemp, straw, peanut shells and perhaps other plants can serve as fiber sources. Paper can be recycled basically the same way it is made.[61] With paper available, artists may desire paint. Paint can be made with mixtures of flour, salt, water, sometimes egg yolks, vegetable dyes and minerals like sulfur for yellow, iron oxide for rust, cobalt for blue. Milk protein can also be reacted with lime (CaO) to make paint. Many DIY paint making websites exist.

Cotton, flax and hemp can be cultivated to make clothing worn next to the skin. When clothes wear out they can be ripped and recycled and even composted. Basalt fiber cloth is actually rather smooth unlike glass fiber and can be used to make outerwear. There are other sources of materials for clothing, shoes and accessories. A completely vegan substitute for leather can be made from fungus mycelium grown on agricultural wastes and byproducts called Mylo.™ [62]

Bolt Threads, the company that makes Mylo™ also makes fibers from spider silk protein obtained from genetically programmed yeast called Microsilk.™ [63] Vats of fungal mycelium and yeast will free up space farm area that would otherwise be used for cotton, hemp, flax, etc. This will also be far more practical than herding cattle for hides in space or on the Moon and Mars.

Bamboo is a useful fast growing crop. Some species can grow 36 inches in 24 hours. Bamboo can be used to reinforce concrete. It has been used traditionally for medicine in Asia and it can be used instead of wood to build houses and schools. Bamboo shoots can feed people and animals. It can be used to make furniture, rugs, toys, kitchen utensils, beer and musical instruments like flutes and drums. Bamboo can make flooring, writing surfaces, pulp for paper making, fishing poles, and filters that can remove salt from seawater. [64] Such filters might be used in reverse osmosis pumps that work to control water salinity in Closed Ecological Life Support Systems or CELSS.

It has been claimed that bamboo can make cloth for bedding, clothing, accessories and diapers. However, according to Wikipedia, the FTC and the Canadian Competition Bureau bamboo textiles are actually rayon made from bamboo treated with harsh chemicals. Bamboo textiles are not made of natural fibers, but bamboo is a good source of cellulose for rayon if that is desired.[65]

Many drugs and medicines can be made from plants. Atropine, codeine, cocaine, caffeine, digitalis, ephedrine, quinine, morphine, reserpine and many other drugs/medicines come from plants.[66] Extracts can be made with water, ethanol or glycerin.

One of the most sought after biomaterials will be ethanol in the form of beer, wine and liqour. Pure concentrated ethanol can be used as an antiseptic and a solvent. It's just a matter of fermentation and distillation.

Toxic chemicals in a closed environment must be avoided. Aerosol spray cans are out. Roll ons and pump sprays for things like deodorants and perfumes could be used exclusively. Perfumes can be made from essential oils from roses, lavender, jasmine, carnations, chamomile, etc. These oils can be extracted with water, ethanol or vegetable oil. Many DIY natural perfume making websites exist.

Vegetable dyes can also be used for clothing. Red cabbages will make purple, onions will make yellow/orange, coffee grounds will make brown, and strawberries will make pink. Salt or vinegar (dilute acetic acid) can be used as fixatives (mordants).[67,68]

Chapter 8: Lunar Cement

Cement Defined

Portland cement is more than just a mixture of lime, CaO, sand and some gypsum (calcium sulfate). Plain mortar for bricks is made by mixing sand with slaked lime (CaOH). This material will absorb CO_2 from the air and become harder. A better mortar is made by mixing Portland cement with sand. Cement is composed of a mixture of calcium silicates, mostly Ca_2SiO_4, Ca_3SiO_5, and calcium aluminate, $Ca_3Al_2O_6$. The calcium aluminate hydrolyzes when mixed with water and forms calcium hydroxide and aluminum hydroxide. The calcium silicates react with the hydroxides and form intermeshed crystals of calcium aluminosilicates.[69]

Concrete is made by mixing cement, sand and crushed rock or gravel with water. The sand and gravel are called aggregate and various other things like ash from coal plants and crushed glass have been used. Less cement is needed for a construction job when using concrete.

Making Lunar Cement & Concrete

On the Moon we will get a mixture of silica, SiO_2, and calcium sulfate, $CaSO_4$, after leaching magnetically beneficiated regolith that has had the iron and ilmenite ($FeTiO_3$ which also contains the titanium) removed, with sulfuric acid to extract aluminum sulfate and magnesium sulfate and some trace metals. Sulfur is fairly abundant in regolith at 500 ppm or more and could be obtained during volatile harvesting. Most of the sulfur and sulfuric acid will be recycled. The sulfates will be dewatered and roasted at 1000 C. or hotter to form oxides and gaseous SO_2 and SO_3 which can be reacted with water to reform sulfuric acid. The $CaSO_4$ can be calcined, or heated to high temperatures (~1500 C.) in solar furnaces, to get lime, CaO. This can be mixed with aluminous Highland regolith in the right proportions and heated in solar furnaces to make small glassy marbles called "clinker" which are then ground in a ball mill. This powder is then mixed with $CaSO_4$ about 5% and ground some more to make cement.

Acid leaching puts precious water at risk and requires complex apparatus to leach and recycle everything, but it's the only way to get $CaSO_4$ on the Moon. Calcining this to get lime might not be the best way to make the actual cement. Anorthite, $CaAl_2Si_2O_8$, which composes 75% to 98% of highland regolith based on analysis of Apollo samples can be roasted at 2000 C. in the vacuum to decompose the mineral and drive off the SiO_2 component leaving calcium aluminate, $CaAl_2O_4$, behind.[70] Calcium aluminate makes good cement. Calcium sulfate is added to retard setting time.

On Earth limestone and clay are mixed together and heated to 1500 C. to get clinker. The limestone breaks down into lime and CO_2. The clay, which is basically just feldspar or plagioclase like the Highland regolith, reacts with the lime to make the mixture of calcium silicates and calcium aluminate called Portland cement.

On the Moon, water will come from polar ice, hydrogen from volatile harvesting combined with oxygen, and imported hydrogen combined with lunar oxygen. Since water is only 1/9 hydrogen, importation could be practical. Concrete made by mixing cement with lunar regolith and gravel will be very useful for construction within pressurized lava tubes someday. It cannot be used out-vac because the precious water will just evaporate into the vacuum and the chemical reactions that make the cement harden will not occur. Concrete items could be cast inside pressurized modules then moved outside. Cement board for walls inside of modules could be made indoors. It might even be possible to spray concrete inside of inflated forms to make habitat.

Concrete can be made from 1 part cement, 2 parts sand (raw regolith that has been sized by screening and sieving), and 4 parts gravel.[71] Thus, a meager amount of cement makes seven times as much concrete. Concrete floors inside pressurized inflatables will provide a sturdy base for heavy machines like large engine lathes weighing 24,000 pounds on Earth (4,000 lbs. on the Moon) and heavy parts like multi-ton steel rolls for rolling mills. It will also resist damage by spilled molten metals from small part casting.

Sulfur Cement

Sulfur might also be used in place of water to make sulfur cement. This is made simply by mixing molten sulfur with sand. No lime is required. Sulfur cement can be mixed with gravel to make concrete.[72] Sulfur can be obtained by roasting large tonnages of regolith at 900 to 1200 C. It has one drawback-- it will melt in the heat of lunar day. Large foil heat shields will be required and we could cast sulfur cement structures in the shadow, perhaps with contour crafting gantries, to make structures that would then be covered with several meters of regolith for thermal insulation and radiation shielding for habitat stationed within the structures.

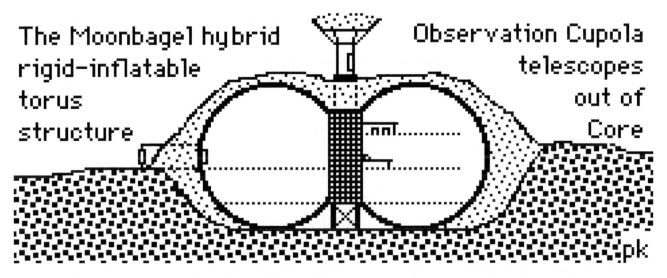

The Moonbagel hybrid rigid-inflatable torus structure

Observation Cupola telescopes out of Core

pk

Fig.18 Pressurized inflatable structure for making molds, machining metals and pouring concrete with sturdy concrete floor. Courtesy of Peter Kokh.

The View Toward the Future

On Mars, mortar will be used inside of pressurized habitat to bond bricks made from martian regolith. We can also mine gypsum, $CaSO_4*H_2O$, on Mars. On the Moon we will need to use mortar made from cement and sand which does not require CO_2 to harden to bond bricks made of sintered basalt perhaps in lava tubes. We will not want mortar to absorb CO_2 from the air of lava tube habitations because carbon is scarce on the Moon and we need it for supporting crops.

Leaching aluminous Highland regolith with sulfuric acid will be a job for more advanced Moon communities when we can manufacture the necessary equipment on the Moon and have water from polar ice mining and water made by combining hydrogen from solar wind implanted volatiles mining with LUNOX in abundance. We will need quite a bit of industry there to do the job. The original "seed packages" of robotic devices will grow and grow using lunar materials to make more equipment until we are ready to seal, pressurize and inhabit lava tubes. Inside the lava tubes, towns for several thousand people will be built with concrete, bricks, plaster, glass and metal mainly iron. There will be no intrusion by ground water on the Moon to erode or rust our structures. These underground towns will be more homey than the metal and inflatable plastic bases we build in the early days of lunar industrialization. They will have gardens and farm sections illuminated by light piped in from the surface during the long day and super efficient microwave sulfur lamps that mimic the spectrum of the Sun without the UV and IR by night.

The sulfuric acid leaching tanks and related equipment will be made with an alloy of abundant lunar iron and about 15% silicon called *duriron* instead of stainless steel. [73] Cast basalt also resists concentrated sulfuric acid and caustic bases like sodium hydroxide, so acid handling equipment might be made with cast basalt. Acid leaching will also produce plenty of silica for glass and calcium sulfate which is dry plaster. Plaster can be wetted and applied between two layers of glass fiber or basalt fiber cloth and allowed to harden to make a wallboard that resists moisture and mildew. We will not need precious paper for wallboard.

Prior to acid leaching, anorthosite must be melted and quenched to break down its crystalline structure. It must be melted to form an amorphous glass then cooled rapidly to prevent recrystallization. Since water cannot be wasted it might be possible to use liquid oxygen. Then the material can be broken up with hammers and ground fine in rod and ball mills. Liquid oxygen would never be used to quench hot metal because it would oxidize the metal, maybe even ignite it, but anorthosite is already oxidized so this might work.

Sulfur is present in regolith in the form of meteoric troilite, FeS, and perhaps other forms. The lunar dust roaster and all isotope separators can get the sulfur along with other elements present in mere parts per million. It's also possible to just roast the regolith at over 900 C. and sulfur will evaporate from the regolith. Sulfur will be used for sulfur cement, sulfuric acid and possibly sodium-sulfur batteries. It can be used for polymers too. It might be used to vulcanize synthetic rubber tires for vehicles that operate in pressurized lava tubes. Sulfuric acid can be used to remove scale from metals if that should be a problem. Heat treating metals in salt pots might lead to scale formation. Sulfuric acid is used to make rayon, a semi-synthetic fiber. Carbon disulfide, also used to make rayon, is made by reacting sulfur and carbon in an electric arc furnace. Iron can be reacted with sulfuric acid to make a bluish green iron sulfate salt that might serve as a pigment for paint and ink. Aluminum sulfate is also called alum and it has numerous industrial uses. Magnesium sulfate is the familiar Epsom salts. Sulfur has other uses. Sulfur dioxide can be used as a refrigerant. However, it is toxic so importation of inert CFC or HFC refrigerants might be wiser. Ultramarine blue, a beautiful pigment, can be made by heating a mixture of sulfur, kaolinite clay and sodium carbonate in a kiln. Clays will not be found on the Moon, but they do exist on Mars. Since ultramarine consists of sulfur, sodium, silicon, aluminum and oxygen, all of which are found on the Moon in good quantities, clever chemists might figure out how to make ultramarine blue with only lunar resources.

Chapter 9: Lunar Tourism

Introduction

Many noble reasons for space industrialization exist. Telecommunication platforms, solar power satellites, intercontinental power relays, helium 3 fusion fuel, microgravity manufacturing of perfect ball bearings, new vaccines and medicines, exotic alloys, precious metals from asteroids, astronomy, SETI, technology spin offs, space tourism and more beckon to us from the space future. Space tourism, particularly lunar tourism, is one of the most enticing rewards of outer space. Life must consist of more than just satisfying animal needs like eating, drinking, getting intoxicated, sleeping and such. There must be more to reach for and experience. Lunar tourism for those few people who could afford it might not seem as noble as providing the world with clean energy from outer space but where would the world be if there were no higher rewards to strive for? Capitalism would die.

Dreams and Nuclear Rockets

Science fiction dreams could come true if reusable nuclear thermal rockets that use liquid hydrogen for reaction mass were perfected and built in substantial numbers in factories the way air liners are built. A single stage atomic rocket with a specific impulse of 850 to 1000 seconds could fly directly from the surface of the Earth to the surface of the Moon. It could refuel on the Moon and return to Earth; a feat that would require aerobraking and a very sophisticated thermal control system like the heat shield tiles of the Space Shuttle. Atomic rockets were not perfected during the space race so less efficient chemical propulsion was used. Unfortunately, research on nuclear rocket propulsion died with the conquest of the Moon and the end of the space race in the sixties. Given the present political climate and public opinions about nuclear power it is highly unlikely that anyone will allow the creation of a lunar tourism industry based on atomic rockets blasting off and landing as frequently as jet airliners because of fear. Flying nuclear reactors would scare the hell out of the public. If an atomic rocket crashed chances are that the reactor would bury itself under several tens of feet of earth and anyone close enough to be exposed to radiation and get cancer, become sterile or have mutated offspring would be killed outright by the impact. However, what if an atomic rocket burns up in mid-air during reentry let's say and large numbers of people are exposed to radioactive material? People are fed up with industrial disasters and many of them believe we are all doomed by climate change. If there is going to be lunar tourism it will have to be based on something other than nuclear thermal rockets.

Propellant from Space

The first step on the long road to lunar tourism is the creation of a manned reusable space plane that can put a person in low Earth orbit for a few tens of thousands of dollars. It would probably be powered by liquid hydrogen and use a combination of jet engines and rocket motors. Such a vehicle could fly people half-way around the world in less than an hour as well as carry people up to space hotels in low and medium Earth orbit. Space hotels would probably be made of inflatable modules. Small vessels based on inflatables with heat shields could be fueled up in orbit with hydrogen and oxygen from Earth's surface. These vessels could fly around the Moon and aerobrake into Earth orbit upon return. This is how lunar tourism would begin. Overflights will not be enough for some adventurers. Landings might take place in open cockpit vehicles.

Growth of the industry will eventually require more propellant obtained on the Moon and from near Earth asteroids. High thrust rockets will be needed to travel from low Earth orbit to a space station at Earth-Moon Lagrange point one in just a few days time. Solar electric spacecraft would spend weeks even months spiraling out to the Moon. Although such craft would use very little reaction mass, life support constraints and radiation exposure in the Van Allen Belts would make it impossible to use low thrust electric propulsion with manned vessels. Cargoes would be sent to the Moon and bases would be "bootstrapped" using local resources. It is not likely that the Moon will be colonized and industrialized for tourism alone. It is more likely that lunar resources will be tapped for the construction of solar power satellites and/or helium 3 mining. Tourism would ride in on the coat tails of the space energy industry. Mass drivers on the Moon capable of launching millions of tons of raw material into space every year could supply propellant to depots at Earth Moon Lagrange pt. 1 and in LEO. Hydrogen and oxygen would come from lunar polar ices at first and later from asteroids.

The infrastructure needed to support a commercial space travel industry will be very impressive. A multitude of lunar industrial settlements where humans and robots work together to extract metals and oxygen from Moon rocks and regolith will be needed along with bases on polar crater rims that send robots down into the darkness of the crater bottoms to mine for ice. Dozens of near Earth asteroids will be mined for water, carbon compounds, metals and oxygen. Large tankers will move mammoth quantities of rocket fuel and oxidizer through space.

To get an idea of how much propellant will be needed, let's envision a thousand spaceships each with a mass of one hundred tons empty and able to carry one hundred passengers. They would take three days to reach the Moon, or more accurately the transportation hub station at EML1, spend twelve hours refueling and being checked out, then travel back to LEO in three days where another twelve hours is devoted to servicing them. This means one round trip per week per ship. If these ships spend two weeks of every year in "dry dock" so that they can be refurbished with new engines and other equipment, they can each make fifty flights per year. One thousand of them could move five million tourists every year. That's about as many people that travel by air every day globally at the present time.

If these ships use nuclear thermal rocket engines, a risk that might be acceptable if the ships never enter the Earth's atmosphere, with a specific impulse of 1000 seconds and the delta V from LEO to L1 is 4100 meters per second then 52 tons of LH_2 will be required for one leg of the flight. If chemical propulsion using LH_2 and LOX is used, 153 tons of propellant will be needed. If silane and LOX with a specific impulse of 340 seconds is used then 242 tons of propellant are required. *The interesting thing about this is that chemical propulsion will use more propellant overall but less hydrogen than nuclear propulsion will.* With LH_2 and LOX, an Isp of 450 seconds, and a six to one oxidizer to fuel ratio, 22 tons of hydrogen will be necessary. With silane only ten tons of hydrogen, less than half as much, is needed as with hydrogen and LOX and only a fifth as much as is needed with nuclear.

Propellant Demands for 100 tons ship and a delta V of 4100 m/s

Nuclear rocket Isp 1000 sec. Exhaust velocity = 0.0098(1000) = 9.8 km/sec
$$e^{(4.1/9.8)} = 1.52$$

152/100 = 1.52   152-100= 52 tons LH_2

LH_2/LOX rocket Isp 450 sec. Exhaust velocity = 4.41 km/sec. $e^{(4.1/4.41)}$ = 2.53   253/100=2.53

253-100= 153 tons propellant using a 1:6 fuel/oxidizer mixture 153/7 = 22 tons LH_2

SiH_4/LOX rocket Isp 340 sec. Exhaust velocity = 3.332 km/sec $e^{(4.1/3.332)}$ = 3.42   342/100=3.42

342-100=242 tons propellant

$SiH_4 + 2O_2 ==> SiO_2 + 2H_2O$ Si = 28 $2H_2$ = 4 $2O_2$=64

28+4+64=96   4/96 X 242 = 10.08 tons hydrogen

It is likely that hydrogen will be the "pinch point." Silicon and oxygen are abundant in Moon rocks, regolith, and C and S-type asteroids. There is enough solar wind implanted hydrogen on the Moon to provide water for early Moon bases but not enough to fuel rockets. There are huge amounts of hydrogen in polar ices but what will it cost to mine that ice? And do we want to waste such a precious resource that could be of immense value to future lunar civilization? It seems the smart thing to do would be to use less powerful silane and LOX to conserve hydrogen. This also eliminates nuclear dangers. Even greater efficiency might be had if silane is used as a carrier fluid for metal powder fuels (aluminum, magnesium, calcium and/or ferrosilicon) in bipropellant rockets.

A fleet of a thousand rocket ships with nuclear thermal motors would use 5,200,000 of hydrogen every year. With LH_2 and LOX propulsion they would use 15,300,000 tons of propellant overall but only 2,185,000 tons of hydrogen. With silane and LOX they would use a total of 24,200,000 tons of propellant but only one million tons of hydrogen. This will demand a lot of mining and material processing and transportation.

Annual Propellant and Hydrogen Demands for 1000 Ship Fleet

1000 ships X 100 flights (50 round trips) X 52 tons LH2 = 5,200,000 tons hydrogen

1000 X 100 X 153 = 15,300,000 tons propellant 1/7 of that is 2,185,000 tons hydrogen

1000 X 100 X 242 = 24,200,000 tons propellant 4/96 of that is 1,000,000 tons hydrogen

Carbonaceous chondrite asteroids contain 3 to 22% water.[74] If an asteroid that is 10% water is found then ninety million tons of asteroidal material must be dug and processed every year to provide one million tons of hydrogen for silane. If that material contains oxygen and silicon in percentages similar to the Moon then there will be more than enough oxygen and silicon for propellant. A quantity of rock several hundred meters wide would have to be mined. In ten years a whole mountain of an asteroid would be mined up. This task would be daunting.

Advanced Propulsion Systems

Fusion drives with specific impulses in the hundreds of thousands versus hundreds for chemical propulsion would use only comparatively tiny amounts of hydrogen for reaction mass. If controlled fusion turns out to be more practical than mining massive amounts of asteroidal and lunar material then fusion could be the way to make the space tourism business practical and profitable. There is another form of propulsion worth considering. Beamed power could make passenger ships with electric drives feasible. Ion drives can be clustered together to get more thrust but the real problem is that the solar or nuclear power plants become prohibitively massive. If a passenger ship is equipped with banks of electric thrusters using hydrogen or sodium vapor for reaction mass and a lightweight receiving antenna to capture power from large multi-gigawatt solar power satellites then high thrust electric propulsion that uses only a fraction of as much propellant as does nuclear thermal or chemical rocket propulsion would be possible. A fusion miracle might not be needed.

The Future

It would be very disheartening if at the end of this century the Moon is no more popular than Antarctica is today with a few tens of thousands of visitors every year. Most of us who were young during the Space Race looked forward to a 21st century with regular and frequent flights to the Moon. If it does become possible for five million tourists to visit the Moon every year, something that will require substantial habitable volume on the Moon perhaps in pressurized lava tubes, then over the course of a 75 year lifetime as many people as there are in the USA today could visit the Moon. It is hard to imagine lunar tourism for all of the Earth's nine to twelve billion people expected by mid-century. Spaceplanes the size of jumbo jets or larger and passenger liners the size of the Battlestar Galactica would be called for. Supplying the liners with propellant from asteroids would be a job comparable to the global coal mining industry today. Fusion or beamed propulsion start looking even better.

It is also possible that in another century or more super materials will make it possible to build space elevators that can haul the masses into space as easily as we ride buses or jets today. Perhaps carbon from the Moon and NEOs could be smelted into graphene with solar furnaces at GEO space stations to make 100,000 km long tethers for space elevators. Ships could be flung off the ends of the space elevator counterweights with no fuel and oxidizer at all. These ships could head towards the Moon or they could rendezvous with cycling stations to Mars. While a six month trip might not deter those with a passion for Mars it could someday be possible for tourists to reach Mars in just 39 days in ships with nuclear electric propulsion. Perhaps space elevators could be used to give slowly accelerating NEP ships a boost on the way to Mars or other worlds of the solar system.

NEOs: Valuable Rocks

Spaceships traveling from LEO to a spaceport at Earth Moon Lagrange point one (EML1) will probably be filled with propellant derived from near Earth objects someday. Industrial bases on the Moon will be "bootstrapped" from a minimal mass of machinery that makes maximum use of on-site materials to self replicate and build mines and mass drivers. Lunar materials would be used to build space shipyards perhaps at L5 where artificially intelligent robotic asteroid mining ships and tankers are built.

Carbonaceous chondrite asteroids can be over 20% water and a few percent organics in material resembling kerogen. The stony component of these asteroids consists mostly of oxygen, silicon, iron, magnesium, small amounts of aluminum and calcium, and numerous trace elements. Once again, there is access to lots of silicon to make silane and extend hydrogen supplies. Asteroidal material would be mined, crushed up, and roasted with solar heat to drive off water and organic compounds. This dried out material could be treated with fluorine to displace oxygen in minerals and form tetrafluorosilane gas (SiF_4, b.p. minus 86 C.). Oxygen and SiF_4 could be separated with membranes. The SiF4 would be decomposed at 850 C. with solar heat to get silicon and recover fluorine. Alternatively, something like a giant mass spectrometer could separate all elements in the asteroid.

All that useless rock that was sought after for its water and organics now becomes valuable property. It becomes a source of oxygen for breathing and propellant and a source of silicon for silane and solar panels. Silica, SiO_2, the main component of glass, obtained by re-oxidizing silicon or boiling it directly out of minerals in the vacuum with focused solar rays, can be used for construction in space. Glass fibers have more tensile strength than steel. Fiberglass made with a polymer matrix from asteroidal organics could find many uses. Iron can be combined with carbon via the ancient crucible steel method. Rods of iron are packed in carbon powder and brought up to red heat for a few days with free solar energy. Carbon dissolves into the iron and forms steel. Space stations and space colonies could be built with steel. Magnesium is a respectable metal compared to unalloyed aluminum and it will not catch fire in the vacuum of outer space. It is a good reflector. Sheets and foils of magnesium could concentrate solar energy onto silicon photovoltaics and increase their power output. It would also be wise to consider slurry fuels made of silane, iron and/or magnesium powders. Magnesium is shock sensitive in liquid oxygen and will detonate. Tanks of magnesium and liquid oxygen slurries might be ignited with electric sparks for blasting into asteroids.

Transporting all this material mined by unpaid robots that never sleep through space will be a challenge. Solar electric propulsion is very efficient and requires only meager amounts of reaction mass. Solar sails many kilometers in diameter or magnetic sails like those envisioned by Andrews and Zubrin might move mountains of raw material through space with no propellant at all.[75] Finally, one must wonder if it will be more profitable to mine small numbers of large asteroids over a kilometer wide and just take the materials wanted and leave the slag behind, or bag and capture large numbers of small asteroids just tens of meters wide and use every last bit of them for propellant and space construction materials in Earth-Moon space?

Requirements for Lunar Tourism

Vacationing on the Moon is a dream that will someday become reality, but making it so will not be easy. Transportation to the Moon is only part of the challenge. A substantial workforce in space consisting of stewards and stewardesses, pilots, technicians, engineers, construction workers, robot teleoperators, drivers, skilled craftsmen, artists, musicians, chefs, doctors, teachers and others needed for a viable civilization as well as large numbers of tourists and business travelers must be kept alive. Humans will require life support, comfort and entertainment. Some wealthier visitors may demand luxury as well. Early trips to the Moon might consist of a single loop around the Moon without landing and return to Earth via free return trajectory. That would be followed by trips that involve several lunar orbits and finally landings on the Moon. Rocketing to the Moon will be expensive so travelers deserve to see as much of the lunar surface as they can during their trip. There must be numerous places to stay in scenic locations and transportation systems on the Moon's surface. What else is required?

Low cost access to LEO by reusable VTOL rocket or HOTOL space planes.

Reusable space ships for travel from LEO to EML1 or LLO space stations.

Reusable landers for travel between EML1 or LLO space stations and lunar surface.

Habitations on the lunar surface. Villages, towns and cities. Roads and railways.

LEO propellant depots and space cryogenic liquid transfer technology.

EML1 or LLO propellant depots with space cryogenic liquid transfer technology.

Mass driver base on lunar surface to launch regolith.

Mass catchers to receive lunar regolith and move it thru space.

Regolith smelting space stations for production of propellants including LOX and metals. Also oxygen production for life support.

Lunar surface propellant production for landers.

Space shipyards to build ships. Stations to service ships.

Factories and garages on the Moon to build and service landers.

Space stations in LEO to serve as ports for transferring passengers and cargo. Will require baggage handling systems and be combined with propellant depots.

Space stations at EML1 or in LLO to serve as ports for transferring passengers and cargo. Will require baggage handling systems and be combined with depots.

Food production in space. Requires large space stations and/or space settlements like Kalpana 2. These stations will also grow crops for fiber, biodegradable plastic, paper, etc. Food packaging and laundry facilities located at stations and settlements. Human waste from ships including CO_2 could be processed by space station/settlement closed ecological life support systems and hydroponic farms. All food and water withdrawn from space settlements by ships must be returned as waste to the closed bio-system loop.

Water production from lunar oxygen, solar wind implanted hydrogen and polar ices.

Stations and/or settlements to house space workers and ship crews.

Shops and small factories in space stations and settlements for making and repairing necessary items for workers as well as travelers like tableware, containers, clothing, bedding, furniture, appliances, machinery parts, souvenirs, sundries, toiletries, medical and office supplies,etc.

Shops and factories on the Moon for making and repairing necessary items.

Lunar ground vehicles and sub-orbital vehicles. Vehicle manufacturing shops/factories and service garages on the Moon.

Telecommunication systems in space linking ships with space stations, lunar surface dwellings and Earth. Land lines and cell towers on the Moon.

Navigation satellite constellations in HEO and polar lunar orbit.

Computerized billing and banking systems. Ticket and reservation agents, etc.

Police forces, security cameras and alarm systems, government buildings, courts, brigs, etc.

Stores and kiosks in stations, settlements and lunar habitations where all sorts of products and services like hair cutting are available for sale to space workers and tourists.

Restaurants, bars, hotels, entertainment venues, offices on the Moon and in stations/settlements.

Medical and dental facilities.

Schools eventually, and libraries. Workers will live in space or the Moon with their families.

Chapels, churches, temples, mosques eventually.

Museums. Historical sites like Apollo landing sites.

Casinos

Sports complexes and gyms/workout centers. Basketball, volleyball, tennis etc. on the Moon.

Dance halls/ballrooms, banquet rooms, theaters (live and movie),etc.

Spacesuits and life support backpacks for people of all shapes and sizes must be available on the Moon, in space and on Earth. Travelers might wear spacesuits during flight to LEO.

The requirements for lunar tourism stand in stark contrast with Earth orbital tourism that can be experienced with just low cost access to LEO in the form of reusable space planes or rockets and re-entry vehicles. Space stations and space settlements could allow lengthy vacations and even permanent living in orbit.

Space Settlement Must Exist

Extensive industry in space and on the Moon must precede all this. Solar power satellite and helium 3 industries are most likely candidates for motivating space industrialization. Asteroid mining for precious metals is another potential profit making activity in space. Earth orbital tourism and pricey real estate in Earth orbiting settlements like Kalpana 2 could whet the appetites of entrepreneurs.

Lunar tourism cannot exist without an extensive amount of space settlement. Food must be produced in space. Launching food from Earth's surface would be ridiculously expensive. Food might be produced on the Moon and launched into space with mass drivers or rockets, but once it is in lunar orbit or at an EML1 (Earth-Moon Lagrange Point One) space station it must be hauled down to Earth orbit to be loaded on ships bound for the Moon. Solar electric tugs might take a year to do that and much of the food might spoil. Perhaps spaceships that travel between LEO and EML1 space stations will load up on enough food from the Moon for an entire round trip when docked at EML1. Food will become waste, including CO_2, and this will have to be recycled given the low supply of organic materials on the Moon. Rockets would be needed to carry all the waste down into the Moon's gravity well. It should be far more efficient to produce the food and recycle wastes in space farms. Robots can do farm work, but humans will also be needed to tend the crops and tend the robots, so livable dwellings must be built for them. A trip to the Moon will take anywhere from about 30 to 80 hours.[76] Unless the ships have centrifuges it is unlikely that there will be any cooking from scratch. People at the space farms can pre-cook the food and package it in foil and corn starch derived compostable plastic containers. Aluminum foil containers can be reheated aboard ships in ovens that use radiant heat. Forced convection ovens with fans could also be used since natural convection doesn't happen in weightlessness. Corn starch bioplastic, PLA, is not very heat resistant but it is easy to recycle or compost. Depending on the number of lunar travelers every year, it is easy to picture large space farm stations and thousands of people engaged in space farming.

It is not technically feasible to build a chemically propelled ship that can lift off from Earth's surface and travel directly to the Moon. Aerodynamic vehicles must be built specifically for traveling from Earth's surface to low Earth orbit and back. Ships that travel between LEO and EML1 have to be lightweight to reduce propellant consumption but large enough to keep travelers comfortable for a good deal of time. Landers can be much smaller because travelers will not spend too much time in them. Three different vehicles each designed for its specific role in transporting tourists to and from the Moon are needed. Space stations in LEO and EML1 where ships dock and passengers and their luggage are transferred from one space vehicle to the next must exist.

Pilots who fly space planes up to LEO can come home every day. Moon lander pilots would live on the Moon. Ship crews would spend long periods of time in space because landing them on Earth or the Moon and returning them to space after every flight would be too costly. There have to be accommodations for crews on lay-overs at the LEO and EML1 space stations. Workers who load and unload baggage, water and food at space dock stations, space ship service technicians and engineers, space station operating technicians and engineers and all other necessary personnel who can't go back to Earth or the Moon every day need nice homes to. There will have to be sizable rotating space stations to house and feed all these workers.

Spending astronomical sums of money to fly to the Moon won't be very popular if travelers don't have nice hotels, recreation facilities, opportunities to visit several scenic locations and the experience of walking on the Moon in a spacesuit. Lunar developers will probably be preceded by Moon mining companies that build manned bases where regolith is excavated and launched with mass drivers. Parts of the bases will be landed intact and other parts will be made on the Moon with on-site resources. An initial "seed" of machines, habitat, vehicles, supplies and replacement parts that has a mass of a few thousand tons could grow exponentially when all systems are working. The Moon mining companies would earn money by selling millions of tons of regolith to solar power satellite building companies. After decades of powersat construction the market will be saturated and Moon miners will need a new source of revenue. Lunar materials and equipment made from those materials could be sold to lunar hotel and resort developers. Hotels could be constructed with 10 to 20 meter wide cylindrical modules made of extruded fused regolith covered with 5 or 6 meters of regolith for protection from cosmic rays. The modules would be connected with metal tunnels and airlocks. Some modules would house visitors, others would house workers, many modules will contain stores and workshops and the majority of modules will be filled with hydroponic gardens. Small factories could build pressurized buses that drive across the Moon on dirt roads and sub-orbital spacecraft for rapid travel to distant locations. In the long run, high speed railroads will be built to connect the larger resorts and the towns and cities that grow up around them.

Lava Tube Cities

There are lava tubes on the Moon large enough to contain towns and small cities. Entrances could be sealed with dams built of boulders and molten regolith that hardens to seal the tubes. Abundant oxygen from regolith could pressurize the caves. The rock and regolith overburden will provide excellent protection from cosmic rays. A "shirtsleeve" environment would be created inside where workers could lay bricks, pour concrete and weld metals just as they do on Earth. Underground cities connected by surface railways and subways could someday emerge on the Moon. There could even be sub-selene fish ponds, swimming pools, homes to rent, parks with trees, golf courses and outdoor barbecues inside the lava tube towns. Microbreweries, gambling casinos and adult entertainment could be found in some of these communities while others will be more family oriented and feature musical productions, amusement parks and museums. Many workers and permanent residents of the Moon will bring their families. Schools, libraries, hospitals, drug stores and shopping centers will be needed. University towns might evolve in the sub-surface worlds of the Moon.

Lunar development will require an enormous financial investment and business leaders may rely on the "if we build it they will come" philosophy. However, without space transportation and support systems that cost billions of dollars this will not be possible. At the same time, without any place to visit on the Moon there will be no reason to go there. Business leaders will have to collaborate and agree upon standards rather than engage in cut-throat competition for the sake of humanity's future in space.

Earth Orbit First

Innovative technologies will be needed to realize low cost access to space. This might be achieved by the Skylon space plane or the SpaceX ITS (Interplanetary Transport System) also known as the BFR (Big Falcon Rocket). More recently it has become the Super Heavy and the Starship. This rocket will consist of a large booster with a winged rocket second stage. This author proposes the development of a rocket based on a modified Shuttle external tank, reusable aerospike main engine module and chemically propelled booster; perhaps the first stage of the Super Heavy or a variant of it. The booster would land on a barge at sea and the main engine module would travel once around the Earth, re-enter and parachute down to Earth near the launch site. The external tank would ride to orbit along with the cargo. It would be used in space along with the cargo module and any packaging. The external tank could supply roughly 30 tons of aluminum alloy, titanium and some polyurethane to space construction enterprises. The cargo module and packaging could supply composites, polyurethane foam and even cardboard. The polyurethane might be broken down into carbon and hydrogen. Nothing would go to waste; therefore this system might be economical especially when assembly line production of external tanks, cargo modules, boosters and main engines is applied. Hulls for lunar ships could be based on external tanks. These would be strong and lightweight. If tourists are launched with this rocket in a large capsule that sits atop the external tank a large number of tanks would be orbited. Hundreds, thousands, even tens of thousands and more tons of aluminum and titanium for building space stations and settlements in orbit 500-600 kilometers up could be delivered in the form of external tanks in addition to actual cargo that might have a mass of 100 tons or more with each flight. Finished modules, hard and inflatable, for space stations that house human and robot crews could be launched. Workers would use various machines to dismantle external tanks and process the metals along with actual cargo into space structures.

Orbital hotels, condominiums and time shares could appear in Earth orbit. These celestial pieces of real estate would be self supporting with hydroponic farms and bioreactors to recycle all oxygen, water, food and waste.[77] Crops like cotton and hemp could be grown for fiber to make cloth and paper. Corn could be cultivated to produce PLA, a biodegradable plastic, from corn starch. Algae and yeast along with the inedible stems and leaves from crops could provide livestock feed for real meat, dairy and eggs. Some people will eat liver, sweetbreads, tongues and brains. Undesirable entrails could be turned into pet food. Cats and dogs will join humanity in the settlement of outer space.

At space stations in Earth orbit space ship components finished on the ground and launched into space could be assembled with large tele-robotic arms. Large structures envisioned for outer space like solar power satellites and the hulls of space settlements will consist of simple components that are repeated and assembled. Space ships might eventually be built in space with materials from the Moon and asteroids, but the first ships would be built on the ground due to their complexity. Propellant depots could be made of external tanks with aluminum foil solar shields. Compressors, refrigeration devices, pumps, corrugated metal hoses, solar panels, electric motors and other elaborate parts for propellant depots would be made on the ground and rocketed into space.

Propulsion to Luna and Beyond

Launching propellant for inter-lunar ships from Earth would be too expensive, although some kind of mountainside mass driver might launch tanks of liquid hydrogen or canisters of metal hydrides. Solar power satellite construction in GEO would involve transforming lunar regolith into silicon solar panels, magnesium reflectors and aluminum frames. A substantial amount of by-product oxygen would result. There would also be a lot of iron and calcium of little use for powersat building. There might also be more than enough silicon. Liquid oxygen, iron, calcium and excess silicon from regolith smelting could be transported from GEO to LEO with solar electric robotic tugs that use ion drives or ablation propulsion in addition to electrodynamic tethers. Substantial quantities of residual hydrogen might be available in LEO. Vehicles that can orbit 30 tons are reasonable developments. One hundred passengers would have a mass of about 10 tons. An extra 20 tons of liquid hydrogen could be piggybacked in the vehicles' fuel tanks. When combined with silicon that's enough to make 160 tons of combustible liquid silane. Silane could be used as a carrier liquid for bi-propellant rockets that burn iron and calcium powders with liquid oxygen. These metals would not yield a very high specific impulse; however, a delta velocity of only about 3.4 kilometers per second over Earth orbital speed is needed to send a ship to the Moon in about 30 hours. Aluminum powder has been found to be a reasonable rocket fuel. In time, there could be hundreds of powersats supplying a substantial amount of Earth's energy demands. When no more powersats are needed except for new ones to replace aging satellites, aluminum could be sold for rocket fuel. Eventually there would be asteroid mining and large amounts of oxygen, metals, hydrogen and carbon from Near Earth Objects could be supplied. With high thrust chemical rockets and propellant in Earth orbit it becomes possible to accelerate ships with electric drives to escape velocity in a matter of minutes instead of days or weeks. Electric drives can deliver low thrust for days or weeks and propel ships to Mars at high velocities. One form of electric propulsion, VASIMR (Variable Specific Impulse Magnetoplasmic Rocket), could get a ship to Mars in only 39 days.

Moon mining, solar power satellite construction, lunar helium 3 mining perhaps, orbital tourism, lunar tourism, free space settlement and the settling of Mars can be viewed as parts of an interdependent inter-related system of commercial enterprises. The goal is not merely the creation of high priced travel and real estate for the privileged few, but the creation of a space faring civilization. Employment and mercantile opportunities for small businesses as well as giant corporations will be plentiful in space. Ultimately, populations larger than the carrying capacity of planet Earth will thrive in free space settlements and other worlds of the solar system like Mars and Titan. Given the vast resources of outer space, the return on investment approaches infinity and the galaxy, not merely the sky, is the limit.

Chapter 10: Spaceships

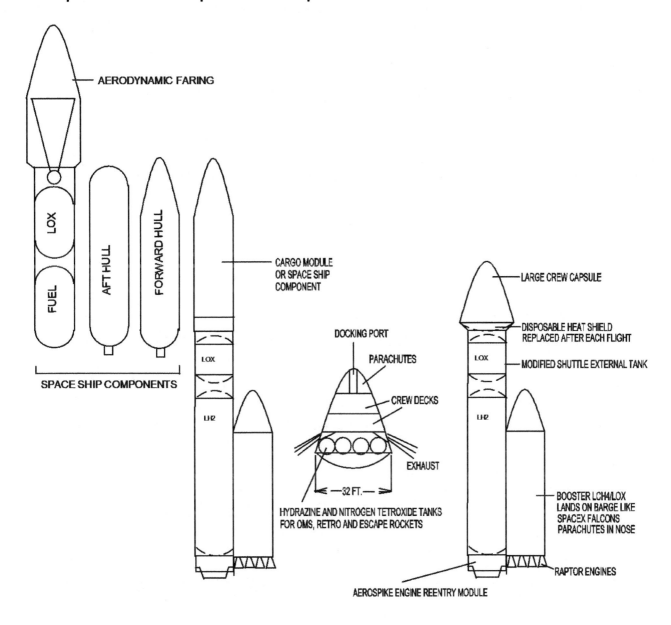

Fig. 19 Rockets to LEO

Shuttle Derived Vehicles

In the seventies, it was suggested that a cargo rocket could be built using a modified Space Shuttle external tank with an aft engine module and four Space Shuttle main engines, an F1 fly back booster using F1 engines like those of the Saturn V rocket first stage burning kerosene and LOX, and a cargo module mounted atop the external tank. It was estimated that this rocket could place 400,000 pounds (181,800 kg.) in low Earth orbit for a space industrialization and space solar power satellite building project.[78] It was never built.

It might be possible to build a similar launcher with more up-to-date technology. Perhaps a booster using F1 engines could be built that lands on a barge like a SpaceX Falcon rocket booster or it could land in a large tungsten steel cable net strung between several barges at sea. This would be much simpler and presumably less costly than building a fly back booster with all the complications of an airplane as well as those of a rocket. SpaceX Merlin engines also burn kerosene and LOX, are very powerful and they are built for reuse. It might be much more effective to use Merlins instead of F1 engines designed back in the sixties. Another possibility is the use of methane and LOX burning SpaceX Raptor engines with the reusable booster.

Reusable Aerospikes

The aft engine module at the bottom of the external tank could be equipped with a low cost disposable heat shield and parachutes (reusable of course) that goes around the Earth once then returns for a soft landing in one of the world's deserts or plains. Reusing those engines should reduce costs especially if refurbishment is done on an assembly line with robots and highly skilled humans. Better yet, instead of using Space Shuttle main engines an aerospike or plug nozzle engine might be constructed. Since the booster just operates at low altitudes conventional bell nozzles could be sufficient. The tank mounted engines must operate from sea level pressure conditions to the vacuum of outer space. An aerospike could be efficient at all altitudes. If the aerospike is effective enough perhaps the external tank could be filled with liquid methane and liquid oxygen instead of liquid hydrogen and liquid oxygen. The tank would have to be modified substantially for this to work. Liquid methane, a soft cryogen, derived from natural gas might be cheaper than liquid hydrogen and involve less capital investment for machinery at the launch pad than would the deep cryogen liquid hydrogen. Aerospike engines are very promising but cooling is difficult. No commercial aerospike engine is in use today. Research and development is badly needed.

Utilizing External Tanks

The external tank would ride to orbit where it would be reused in space. External tanks could be used as spaceship hulls, as spaceship propellant tanks, as parts of space stations, as space propellant depot tanks, as a source of refined metal for construction in space and possibly as metallic powder fuel for spacecraft. It would be far more cost effective to make use of the external tank in space instead of throwing this thirty ton sixty million dollar item away with each rocket launch. Mass production of external tanks on a largely automated assembly line would reduce costs. An aggressive space industrialization program would have to be in full swing with several rocket launches per week to justify this and make it economically viable. There's no use piling up a mountain of hardware if there is no market for it.

It seems that the Space Shuttle was expected to do too much or serve too many roles. The Shuttle was a rocket/airplane hybrid designed for hauling cargo as well as humans. Rockets that are rockets and not combinations with airplanes and all the resultant complications, costs and reduction in reliability are needed. The rocket described above could be fitted with a large space capsule instead of a cargo module for transporting crews into space. A conical capsule about 32 feet wide (roughly ten meters) at the base could carry up to 100 people into low Earth orbit. Its thrusters, orbital maneuvering system, retro-rockets and escape rockets could be powered by space storable hypergolic hydrazine and nitrogen tetroxide. The liquid fuel booster would provide more safety than the solid boosters of the Shuttle did. Solid rockets are hard to control. Once they are ignited they cannot easily be shut down like a liquid fueled rocket can. If a liquid fueled booster breaks loose it can be deactivated immediately and disasters like the Challenger could be averted. With a capsule mounted on top of the stack rather than mounted to the side like the Shuttle orbiter it is possible to fire escape rockets and get away from an exploding rocket beneath. The capsule would have a disposable re-entry heat shield instead of the high maintenance tiles of the Shuttle. Heat shields used on main engine return modules and passenger carrying capsules could be removed, smashed up and recycled after each flight. New shields could be mass produced on an assembly line to reduce costs. After a mission the capsule and crew would parachute down to a soft landing on land in a desert or plain like the main engine module and be trucked inexpensively or shipped by barge to a launch base for refurbishment and reuse.

By making use of the external tank in space this system is essentially 100% reusable. Nothing is wasted. After one hundred missions at least 3,000 english tons (2,727 metric tons) of aluminum, titanium, lithium and copper from external tanks would be placed in LEO along with some polyurethane tank insulation that could serve as a source of hydrogen and carbon. A few tons of residual hydrogen and oxygen in each of the external tanks could also be scavenged. Cargo canisters and packaging could also be a source of metals, plastics and composite materials. This would be in addition to the actual cargoes totaling about 20,000 english tons (18,180 metric tons).

Processing ETs in Space

A space station in LEO composed partly of external tanks would be needed to process cargoes and external tanks. Humans and robots could assemble tanks into clusters and fit them with solar shields, solar panels, batteries or fuel cell systems, pumps, plumbing and boil-off reliquefying machinery to establish propellant depots for manned spacecraft to the Moon or Mars. Space ships made partly of external tanks could be assembled in orbit. Tanks could be disassembled with lasers or high speed cutting tools since aluminum is rather soft. The metals could be powdered or drawn into wires to feed 3D printers for making all sorts of parts. When 3D printing is combined with conventional manufacturing processes like CNC machining, powder metallurgy, rolling and forging most of the components for ships, stations, observatories and fuel depots could be made in space. Plenty of raw material would be available from external tanks.

Certainly, much work could be done in space. Asteroid mining, lunar industrialization for space solar power satellite construction or helium 3 mining, tourism, large telecomm platform building, Mars exploration and settlement, space observatories for studying exo-planets and more could all benefit.

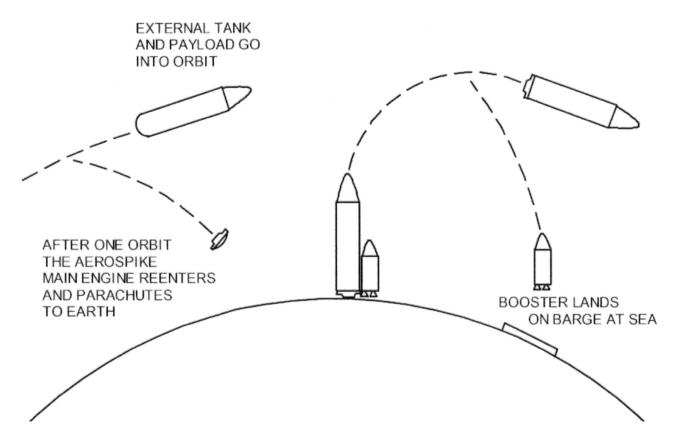

EXTERNAL TANK
AND PAYLOAD GO
INTO ORBIT

AFTER ONE ORBIT
THE AEROSPIKE
MAIN ENGINE REENTERS
AND PARACHUTES
TO EARTH

BOOSTER LANDS
ON BARGE AT SEA

Fig. 20 Booster and aft engine module recovered. ET used in space.

100 passenger reusable Earth orbital capsule

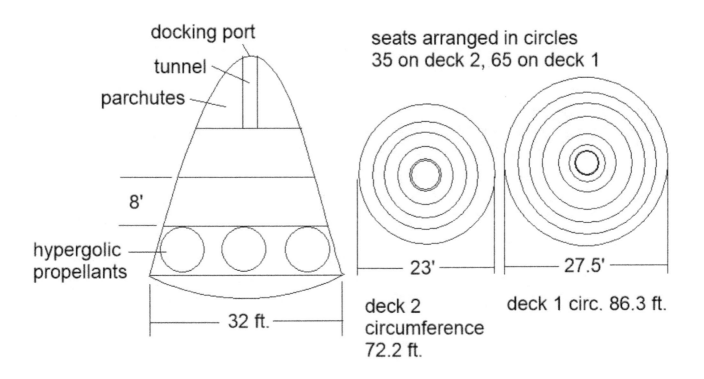

Fig.21 Reusable space capsule for transporting workers and tourists to LEO

Inter-lunar Ships

Lightweight ships for flight from LEO to Earth Moon Lagrange point 1 (EML1) or Low Lunar Orbit (LLO) could be built of hulls based on the Space Shuttle External Tank. The SSET was a prime aerospace vessel that stood 154 ft. tall and 27.5 ft wide, yet it only weighed 66,000 pounds. A modified tank is at the core of NASA's Space Launch System.

Someday there will be hundreds, perhaps thousands, of spaceships hauling passengers to the Moon and other destinations in space. Reusable rockets and space planes will haul people to LEO where they will dock at space stations and transfer to ships that fly to low lunar orbit (LLO) or EML1. There they will dock at space stations and transfer to landers and descend to the lunar surface. Some ships will go to Mars orbit where space stations and landers are waiting. The ships that fly from LEO to the Moon or Mars will be very lightweight and flimsy compared to vehicles that have to climb out of Earth's gravity well. I figure that these spaceships will be so complicated and require so much human and AI robot labor that they should be built on the ground then launched into orbit in modular sections that are then assembled in space I think it will be a long time before space industrial capacity will be great enough to build spaceships from scratch.

Solar power satellites on the other hand are huge but not that complicated. They are just frames, solar panels, wires, amplitrons, a big dish antenna and some attitude control flywheels and thrusters. Most of the parts will be mass produced and repetitive assembly of parts can be done by machines mostly. Some of those machines will be autonomous AI and others will be teleoperated by humans in space and on the ground. There aren't that many separate discrete parts or components for a space solar power satellite (SSPS); maybe just a few dozen different kinds of components. Heat pipes and radiators might be needed, but aluminum heat pipes can be extruded and radiators fins rolled. Making these components will be less complicated than making solar panels. Cobalt-samarium magnets and platinum filaments for the Amplitrons will be required Those would have to come from Earth until we could get cobalt and platinum from lunar meteoric iron fines (eventually asteroids) and samarium from lunar KREEP minerals which contain Rare Earth Elements.

The early space materials and manufacturing industry will be crude but it will evolve to higher complexity with time. We will use the MUS/cle strategy, a term coined by Peter Kokh....MUS = Massive Unitary Simple and cle = complex lightweight electronic or complex lightweight and expensive. MUS parts will be made in space and cle parts will be brought up from Earth

Spaceships could be made of aluminum and titanium alloys like the SSET or they could be made with carbon composite passenger sections and high tensile strength steel for the propellant tanks. Simple pressure feed for the engines will make them reliable, cheap and easily refurbished. High thrusts will not be required in space so the rocket engines for ships that travel from LEO to places beyond will not have to be fancy high stress bearing devices that must endure high pressures and temperatures like the costly Space Shuttle main engines. Thus, they will be good for many firings. They will have highly expanded vacuum nozzles for increased performance.

For pure spaceships that operate only in space we have discussed relatively low thrust pressure fed engines that endure low pressures and temperatures so they have long lifetimes, unlike so many high performance rocket motors that stress materials to their limits and thus have short lifetimes.

Typically, a turbine drives the fuel and oxidizer pumps and drives a hydraulic pump too or there is a separate turbine to drive the hydraulic pump that operates the gimbals for steering the rocket. Silane and oxygen would not be good for a turbine because silica deposits will form on the turbine blades and ruin it. Hydrogen and oxygen or hydrogen peroxide decomposition might be used for turbines, but there is another way. I am suggesting an electric system using electric motors and jack screws to swivel the engine and nozzle for steering the ship. No hot turbines that must be inspected, refurbished or replaced after use and no pressurized hydraulics and valve bodies that might leak or get stuck and malfunction. The motors on the jack screws will be powered by some advanced form of lithium ion battery. Solar panels will keep the batteries charged. One thing I like about batteries over fuel cells is their simplicity by comparison and you don't need to electrolyze the water to recharge the hydrogen and oxygen tanks of the fuel cells. This will all be easier to deal with than the complex high performance rocket engines in use today needed to overcome gravity.

These ships would be a lot like airliners with 200 or more seats on board. They could reach EML1 or LLO in 30 hours. That would be a long time to sit, but weightlessness means no pressure on lower body parts. There could and should be floating rooms with observation domes that passengers can frolic in. Noise proof sleeping cubicles could be provided for people who can't sleep in their seats. Seats should have windows made of polycarbonate several inches thick to shield out solar flare particles. Some ships could have centrifuges that provide 1/6 G and small cabins. This might make travel more comfortable for those who have difficulty with space sickness, space food and zero-G toilets. Ships with centrifuges would be more massive so they might not travel as fast. It might take more like 72 hours to reach the Moon. Not only will these ships be more expensive to build they won't be able to make as many flights every year so they won't make as much money as the faster ships unless ticket prices are increased. You get what you pay for. A weightless flight lasting 30 hours with time for zero-G fun and a lower ticket price might be far more popular than three days in a lunar equivalent Gee cabin. Passengers could still get out of their cabins and visit the floating rooms, but if they get sick they will be trapped in the centrifuge. Space sickness remedies and prophylactics are called for. Centrifuges making lunar equivalent or Mars equivalent "gravity" and private cabins will be a necessity for passenger ships to Mars, but the Moon should be within reach without them.

CHEMICALLY PROPELLED INTERLUNAR PASSENGER SHIP 30 HRS. LEO TO EML1 OR LLO

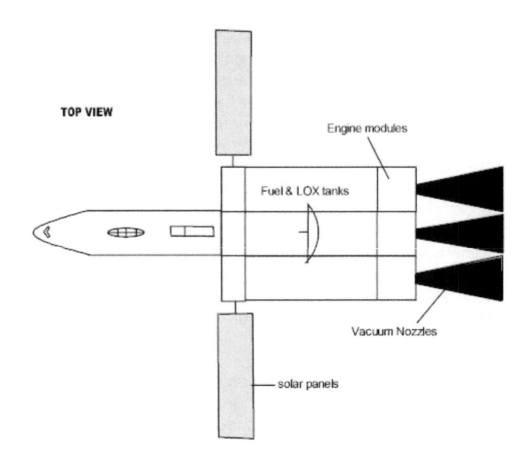

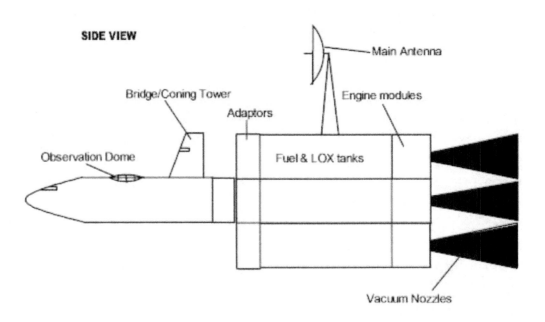

Fig. 22 Inter-lunar ship made of hulls based on Shuttle External Tank

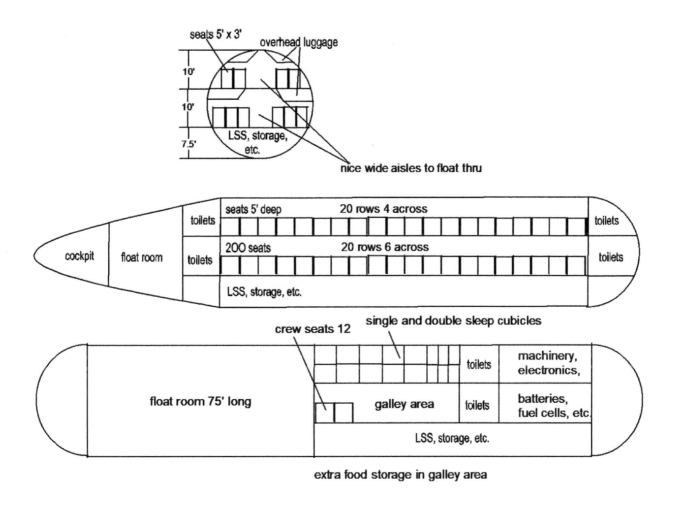

Fig. 23 Interlunar ship seating arrangement

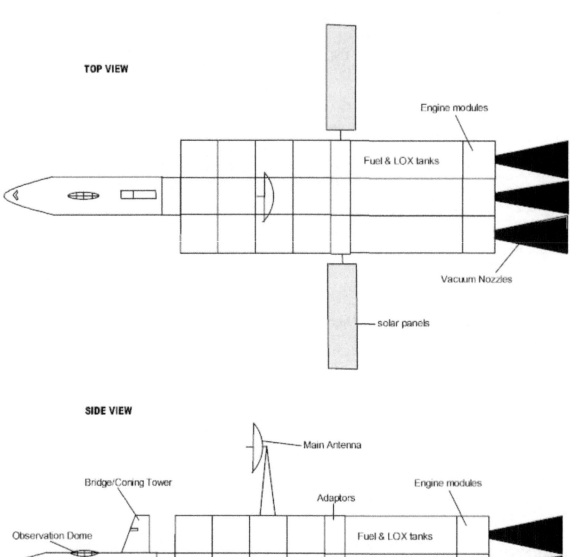

CHEMICALLY PROPELLED INTERLUNAR PASSENGER SHIP 72 HRS. LEO TO EML1 OR LLO

Fig. 24 Ship with centrifuges that rotate only fast enough to produce 1/6 G for passenger comfort.

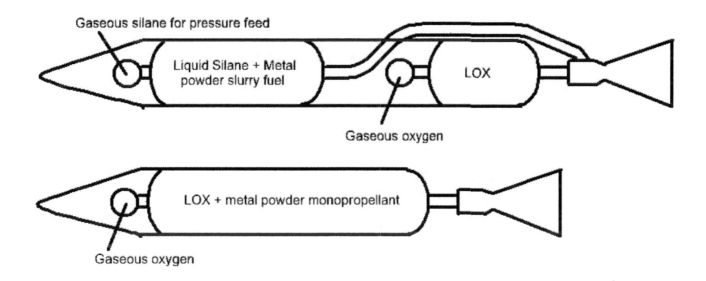

Fig. 25 Pressure fed rockets

A ship from LEO to EML1 and back will have to fire rockets when it leaves LEO and again when it brakes into EML1. It will refuel then rocket away from L1 and retro rocket back into LEO. The ship will have to fire rocket motors 4 times with each round trip. How many firings can it make before motors have to be replaced or overhauled?? I propose simple pressure fed rocket motors. Just valves for the pressurant tanks and valves into the firing chambers but no turbopumps. They will be very simple and thus very reliable, inexpensive and simple to replace or refurbish. Pressure fed rockets aren't used on Earth because the only way to get high fuel/oxidizer flow rates for high thrust that's needed to overcome gravity is to use very high pressures and that means the tanks have to be real heavy. In space you don't need much thrust to accelerate a ship so lower pressures and lighter tanks can do the job. On Earth you need more thrust than the weight of the rocket to get it to lift off. In space thrust equal to the mass of the rocket will produce acceleration at 1G. If thrust is only 1/4 as much of the mass of the ship it will accelerate at about 0.25 G which is enough to reach escape velocity in short notice. Since a ship in LEO needs to add 3.2

km/s to reach escape velocity and 0.25 G equals 2.45 m/s^2 it will take only 22 minutes to achieve escape velocity. During acceleration metal powders suspended in silane would tend to settle out towards the rear of the tank. To prevent this, an agitation system will be needed in the fuel tank. Ullage motors burning silane and oxygen gases will also be needed to move liquid propellants to the rear of tanks before firing main motors.

MAKING SILANE

Hydrogen from Earth, Moon or Asteroids   Chlorine from Earth

Magnesium and Silicon from Moon or Asteroids

$2Mg + Si ==> Mg_2Si$ $2H_2 + 4Cl ==> 4 HCl$

$Mg_2Si + 4HCl ==> SiH_{4\,(g)} + 2\ MgCl_{2\,(s)}$

Silane gas boils off and is liquefied. Magnesium chloride remains solid and is then put through electrolysis to recover and recycle Mg and Cl

Fig. 26 Silane production process

As discussed in the previous chapter, using silane for rocket fuel can cut hydrogen demand in half. If silane can be used as a carrier fluid for powdered metal based slurry fuels, hydrogen consumption can be reduced even more. Hydrogen could be piggybacked into LEO with payloads from Earth, obtained from lunar polar ices and mined from NEOs. Since lunar solar wind implanted volatiles are such a low density resource they will probably be reserved for life support and other uses but not rocket fuel. If helium 3 fusion becomes possible and economical in the future then large quantities of these volatiles will result from helium 3 mining on the Moon. There are millions of tons of ice in lunar polar craters. The exact amount and form it is in (sheets of ice, crystals mixed with regolith?) are unknown presently. Beyond the Moon and NEOs, hydrogen could come from ices of the moons of the outer planets. It could also come from the atmospheres of the Gas Giants. That's a resource humanity would have trouble exhausting! Rockets might be replaced by space elevators someday. Spacecraft could simply be flung off the ends of the tether with no propellant at all.

Chapter 11: Asteroids, Orbital Refueling, Electric Propulsion and Beyond

Return to the Moon

Returning to the Moon with a plan to stay and industrialize it seems likely sometime in this 21st century. Falcon Heavy which can put over 50 metric tons in LEO (low Earth orbit) could be used. Given the shifting winds of politics there is no telling if the Space Launch System for returning to the Moon will ever "get off the ground." SpaceX has also proposed a larger rocket that can orbit well over 100 metric tons for sending humans to Mars. Blue Origin is also working on large rockets. Huge chemically propelled rockets are fine for climbing out of Earth's gravity well, going into orbit and launching small payloads to other planets. Booster stages can be recovered and reused to reduce costs. Upper stages might be refueled in orbit and reused or dismantled for construction materials.

If really large cargoes are to be sent to the Moon and planets of the solar system spacecraft with highly efficient electric drives like electrostatic ion drives, Hall thrusters, plasma thrusters or VASIMR will be needed. These use very little reaction mass or "rocket fuel" compared to chemical or nuclear thermal rockets, but they have low thrusts and take months just to reach the Moon. Slowly spiraling away from Earth means lengthy exposure to Van Allen Belt (VAB) radiation that could be deadly for humans. The VAB radiation could reduce the efficiency of solar panels and damage radiation sensitive cargoes. Some cargoes could be shielded and nuclear power plants might be preferable to solar panels.

High thrust chemical rockets will be needed to send humans to the Moon in just a few days' time and limit time spent in the VABs. Chemical rockets could rapidly accelerate ships in LEO up to escape velocity and those ships could then activate electric drives and reach Mars faster. These rockets could be reused to cut costs. Spent upper stages might be pressed into duty. Perhaps upper stages with SpaceX Raptor engines that use liquid methane (LCH_4) and liquid oxygen (LOX) could do the job. Liquid methane and liquid oxygen are "soft" cryogens unlike LH_2 which is a "deep" cryogen that is more difficult to store in space for long periods of time. Orbital refueling infrastructure in LEO and possibly

EML1(Earth-Moon Lagrange 1) or LLO (low lunar orbit) would be necessary. Propellant could be rocketed up to LEO from Earth and stored in orbital depots. Refueling depots in LLO or at EML1 could be loaded up with propellant from the Moon in the form of hydrogen and oxygen from polar ices or metallic powders and LOX from regolith mined just about anywhere on the Moon. Eventually LEO depots will be filled with propellants from the Moon and/or near Earth asteroids.

Orbital Refueling

Refueling in space is not just a simple matter of connecting a hose between two spacecraft. Cryogenic propellant will be transferred through corgurrated metal hoses like the flex pipe connecting stoves to gas lines. Plastic hoses would freeze up and crack. Spacecraft carrying propellant will have to fire ullage motors to shift the propellants to the bottom of their tanks or they would have to rotate to produce centrifugal force that pushes the rocket fuel to the tank bottom. Gas pressure would fill the tanks to push the propellants out. The fluids would move through the hose to a pump that drives them under pressure into the receiving spacecraft's tanks. As the super cold liquids are injected the warmer empty receiving tanks might cause them to vaporize, so precooling of the hoses, pumps and the receiving tanks will be necessary. This would have to be done behind foil solar shields. Some of the fluid might boil off and it would have to be recaptured and stored in large tanks made from upper stages or external tanks and reliquefied. It is doubtful that one load of propellant rocketed up to orbit will be enough to fuel an interplanetary spaceship. It will be necessary to store up several payloads of fuel and oxidizer in tank farms at orbital refueling depots that have solar shields and boil-off reliquefying machinery. Once enough propellant is stocked up it can be transferred to a spaceship shortly before it leaves LEO. Orbital refueling has yet to be perfected and no space depots exist today. This technology must be developed before any large scale manned exploration and industrialization of the Moon, asteroids and Mars can happen.

Planetary Defense

Asteroid defense should not be neglected in favor of a lunar development program and missions to Mars. The same orbital refueling infrastructure needed for lunar industrialization and Mars exploration programs could be used to fuel robotic asteroid deflection spacecraft. Programs to create protection from potential large impactors, lunar industry and Mars missions should proceed simultaneously. *Orbital refueling infrastructure would be the logical first step for all these objectives.*

A mass driver on the Moon could supply mass for a gravity tractor that could deflect threatening asteroids if there is enough time for warning in advance. Another real "pull yourself up by your bootstraps" idea would be to capture a small asteroid say 10m diameter that can be reached with a delta V of less than 500 m/s and outfit it with some kind of propulsion system and use it as a gravity tractor. It may also be possible to send a robot to a threatening asteroid that digs up some of the asteroid to build up mass for a gravity tractor then start thrusting away. In the worst case a small spacecraft could deliver a nuclear bomb to the oncoming asteroid and detonate it. If this or any other venture beyond Earth orbit is to work there will be a need for nuclear-electric propulsion (NEP) or solar-electric propulsion (SEP) and plenty of reaction mass. Some rather smart AI and weightless excavating equipment would also be required.

EROs: Easily Retrievable Objects

Fuel and oxidizer payloads would be rocketed into LEO at first. Eventually, fuel and oxidizer will come from the Moon and near Earth asteroids. There are at least a dozen near Earth asteroids that can be propelled to Sun Earth L1 (SEL1) or Sun Earth L2 (SEL2) with a velocity change of 500 meters per second or less. These Easily Retrievable Objects (EROs) are about 2 to 20 meters in diameter and there are probably more out there that have yet to be detected because of their small size.[79]

If these EROs are composed of rock, we can foresee based on the study of meteorites that they will probably be mostly silicates containing iron and magnesium. Oxygen will be the major component and oxygen is the major component of rocket propellant. Eight weights of oxygen combine with one weight of hydrogen. The Space Shuttle main engines used a ratio of six weights of LOX for every single weight of LH_2. A methane/LOX burning rocket engine would use about four weights of oxygen for every weight of methane. Metal powder base fuels might also be used. Experiments have shown that aluminum and LOX in a monopropellant slurry make an effective combination with a specific impulse of about 270 seconds. According to Peter Kokh, aluminum powder burns better if it is mixed with calcium powder. Stony asteroids will contain some aluminum and calcium but much more silicon, iron and magnesium. Iron powder could make a low performance fuel. Magnesium powder in LOX has been shown to be shock and vibration sensitive and will detonate.[80] Magnesium might not make a good rocket fuel but it could make a good explosive in the nitrate poor environments of outer space and the Moon. Silicon burns with as much heat per weight unit as aluminum, but there is little or no data on silicon/LOX slurry monopropellants. Silicon can be combined with hydrogen to make silane, SiH_4. Silane burns with oxygen at a mixture ratio of

1:2, but much less hydrogen is needed if silane is used. Some simple stochiometery and the rocket equation can be used to determine that the silane/LOX combination yielding about 340 seconds Isp will require only about half as much hydrogen as the LH_2/LOX combination. Asteroids would supply an abundance of silicon. Manufacturing thin film silicon solar panels and integrated circuits in space won't demand extremely large amounts of silicon. The best thing to do seems to be to send hydrogen up from Earth and combine it with asteroid silicon for silane which is also a "soft" cryogen like LCH_4 and LOX. Iron (steel), magnesium, aluminum and calcium (for electrical conductors) could be reserved for space construction purposes.

Processing Asteroids

Processing asteroids to get useful materials will be a challenge. Electrical and chemical methods for extracting useful elements from lunar regolith and rock could be applied. Solid rock would have to be broken down into a powder for various extraction methods to work. Perhaps a small asteroid could be placed in a huge Kevlar bag. Holes could be drilled into it and explosives planted. The asteroid could be blasted into a pile of rubble. Some asteroids already seem to be floating rubble piles resulting from collision with other asteroids. Those rubble piles could be surrounded with Kevlar bags and spacecraft with electric drives could haul them to SEL1 or SEL2. Rubble piles, natural or man-made, in Kevlar bags could be rotated by applying small amounts of thrust to generate centrifugal force to drive the pieces of rock into jaw crushers and then through several stages of centrifugal grinders to powder them. The powder could then be processed in various ways to get oxygen, silicon and metals. If the asteroid is carbonaceous it will be possible to roast out water and carbon bearing tarry material.

If the asteroid is metallic it will consist of mostly solid iron and nickel with other metals like platinum. These asteroids won't supply oxygen, silicon, magnesium, etc. They will still be very valuable. Iron and nickel could be combined with carbon from C-type asteroids to make steel. This would involve packing iron rods in carbon powder and bringing them up to red heat in solar furnaces for a few days. Asteroid metal could even be processed in something like a huge mass spectrometer to separate all the elements. If this proves to be impractical iron powder or briquettes could be treated with hot carbon monoxide gas to form gaseous carbonyls of iron and nickel that can be piped off and decomposed with intense heat in solar furnaces to get pure iron and nickel and recover CO gas. The residue of metals left after treatment with CO to remove iron and nickel will be enriched in cobalt, platinum and other metals. Breaking down solid iron into smaller pieces and even powders cannot be done with explosives, jaw crushers and grinders the way rock can. High energy electric sparks or plasma torches might be used to cut up solid metal. Concentrated solar energy might be used. Lasers seem like the best idea. High power lasers will soon be deployed on aircraft, tanks and ships for weapons. In a decade or two lasers that are up to the job of cutting up metallic asteroids are likely to exist. Pieces of metal can be melted down and sprayed thru nozzles with an inert gas to form powders. Spinning metal rods can be melted with electric sparks and particles will be thrown off by centrifugal force. The particles cool and powder forms. This second method does not need an inert gas that has to be recycled somehow.

Electric Drives for Fuel Efficiency and Speed

Electric propulsion is needed to overcome the "tyranny of the rocket equation." They can reach higher velocities than chemical rockets or they can reach similar velocities with much less reaction mass. Even so, electric drives have drawbacks. They accelerate slowly. The require massive power sources. They endure more gravitational losses than chemical or nuclear thermal rockets do. Not much can be done about gravitational losses. Highly efficient solar panels and nuclear power plants for space are in the works. Beamed propulsion is another possibility. A massive solar power satellite could beam energy to a lightweight receiving antenna on the spacecraft. It might take months to reach the Moon or even Sun-Earth Lagrange regions with electric drives but robotic cargo ships won't be deterred by that. For journeys to Mars, Mercury or the outer solar system spacecraft with electric drives would take quite a while to escape from LEO but they can apply constant low thrust for months and reach higher velocities than chemical rockets to reach distant worlds like Mars or the moons of the Gas Giant planets faster. With fuel and oxidizer from captured asteroids and orbital refueling infrastructure manned spacecraft with electric drives could be rocketed quickly out of LEO and up to escape velocity, spend very little time passing thru the Van Allen Belts, and reach other worlds in the solar system faster than chemical, nuclear thermal or electric propulsion alone.

Asteroid mining, orbital refueling and electric propulsion combined can open the door to the solar system. Intelligent robots will pave the way. Some near Earth asteroids might require less delta V to reach than the Moon, but launch windows to and from these objects can be years apart. Keeping humans alive and sane in space for years without resupply from Mother Earth will be too difficult. Robotic asteroid mining spacecraft will be needed to do the job. Some asteroids contain water and hydrocarbon compounds. They could be a source of hydrogen and carbon as well as oxygen. Lunar polar ices could supply hydrogen and oxygen too. When these asteroidal and lunar materials are available it will no longer be necessary to rocket them up from Earth at great cost. Stony asteroids and the Moon contain an abundance of silicon which can be used with hydrogen to make silane and extend the hydrogen supply. By using powdered silicon, iron, aluminum and possibly calcium and magnesium mixed with silane for slurry fuels even more hydrogen could be conserved. Hydrogen makes the best reaction mass for VASIMR because the highest exhaust velocity and specific impulse can be obtained with it. Other elements from the Moon that might be used as reaction mass for electric drives are sodium or potassium because of their low boiling points and magnesium which boils at just about 200 to 300 degrees C. hotter. Metallic vapors could be ionized or heated to a plasma and expelled into the vacuum of space.

Tapping the resources of the Moon and near Earth asteroids for rocket fuel will dramatically change the scenario in space. Gigantic three stage rockets with comparatively small payloads will be replaced with smaller reusable rockets to LEO and reusable space tugs that operate between Earth orbit and lunar orbit or beyond. The first step will be establishing orbital refueling stations and stocking them up with propellant from Earth to fuel up the great ships of discovery and commerce, robotic as well as manned.

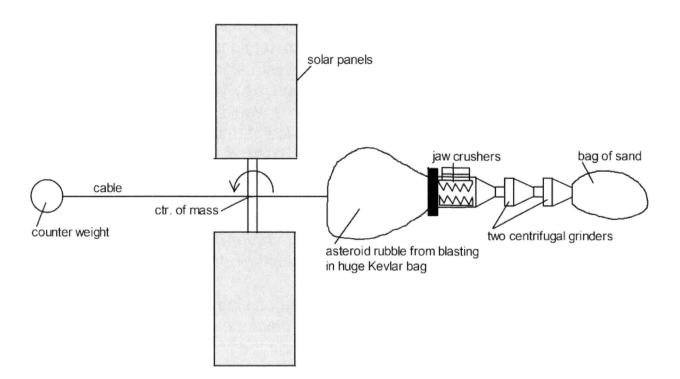

Fig. 27 Rotating system for blasting, crushing and grinding up small asteroids.

Chapter 12: Interplanetary Telecommunication

Two way calls between planets other than Earth and Moon will not be possible due to the time it takes for electromagnetic waves to travel at the speed of light. Depending on it's distance, Mars which ranges from 33.9 million to 249 million miles away, it will take 3 to 22 minutes for a message sent by laser or radio to get there from the Earth, Earth orbit or the Moon.[81] If you send a message you will have to wait at least 6 to 44 minutes for a reply. Between here and Saturn the wait would be several hours. Communicating with the Kuiper Belt will take a couple of days or weeks. Messaging the Oort cloud could take months even years. It looks like we will send video messages, voice mails, emails, faxes and text messages to other worlds and just wait for a reply. Behind this magical service will be an incredible and complicated mass of infrastructure.

Phone and internet users will connect via towers or land lines to the incredible international and someday interplanetary telephone network. A PSTN—Public Switched Telephone Network or an ISDN—Integrated Services Digital Network, will route your calls to an International Gateway for communicating with other countries if the ground stations are in different nations. In the future we will need an Interplanetary Gateway. The fabulous phone network will get your message routed to a ground station that up-links to one of at least three space stations in GEO. Up-links will be done with microwave radio transceivers because lasers, especially infrared lasers, can be blocked by clouds or absorbed and distorted by the atmosphere and dust. Everyone who has satellite TV knows that in stormy weather service can be interrupted. If weather is blocking access to the space station the network will have to route your message to a ground station where things are sunnier or at least the night air is clear. Messages in digital form will then be transmitted from the space station via laser beam to stations orbiting other worlds. Lasers are much tighter than focused radio wave beams so their energy won't spread out all over space, therefore lasers will be more energy efficient. Also, lasers have more data bandwidth than radio. At the receiving station, telescopes will pick up the laser beam and focus it on a CCD (charged coupled device) that converts light to electricity and those signals will go into the radio system and down to a ground station on the destination planet. The message will be routed to the person you are calling. What if the person is roaming?? Your mobile phone sends out signals and computers store data on which cell your signal is strongest. The computers will direct your message to where ever the person is. If they are outside of any tower range, hopefully they will have a satellite mobile phone and computers will locate them based on GPS data from their phone. Settlements in solar orbit will have their own laser signal receivers pointed at various worlds and their own internal phone network. There will also be relay stations in solar orbit that have lasers for communicating with worlds that are behind the Sun. These will also boost signals on their way to planets and settlements that are very distant. The relay stations will also convey signals to multitudes of ships out in space.

To call another planet to send a voice mail, text, video or fax you will have to "dial" the interplanetary prefix, then the planet code, the country code, area code and seven digit number. To call a ship you might "dial" the interplanetary prefix, then the ship code, and the other person's private number assigned by the ship company. Cell phones might not work in a ship. The metal hull and bulkheads will absorb radio waves like a Faraday cage. It might be possible to have small antennas throughout the ship connected by coaxial cable to the ship's laser transceiver. This would allow travelers to use their wireless phone as easily as they do back home and have wi-fi also. Then it is probably just going to be a matter of using the interplanetary prefix, the ship code, the area code and seven digit number. It might just be easier to send an email and let the internet routers and servers figure out how to convey your message to the other person.

At present, it is possible to call the International Space Station. The number has a Houston area code because that's where the radio up-link network is at the Johnson Space Center.[82] Engineers there can relay audio into the radio network but they don't do that for just anybody.

Communicating with people in LEO will also require ground stations and GEO satellites. As a space station in LEO circles the Earth it is visible on the ground for just minutes. An impossible number of ground stations would be required to keep track of stations in LEO. From a satellite in GEO it is possible to cover about one-third of the Earth's surface and any other satellites or stations traveling over that area. Something in LEO would be visible for about 30 minutes at a time. With three satellites in GEO that can hand off signals to each other as low Earth orbiters pass over the "horizon" and into another GEO satellite's territory it will be possible to track and communicate with any spacecraft in LEO at any time. The GEO satellites might need to handle a lot of data so they are more likely to be large stations, possibly manned, similar to large telephone exchanges. Only three ground stations that uplink to the GEO stations that relay signals from the ground to LEO and back, and signals between spacecraft in LEO, would be needed. Real time two-way communication will be possible since radio waves can travel from Earth's surface to GEO and down to LEO in just a fraction of a second.

INTERPLANETARY TELECOMMUNICATION SYSTEM FOR TEXT, FAX, VOICEMAIL & EMAIL MESSAGES

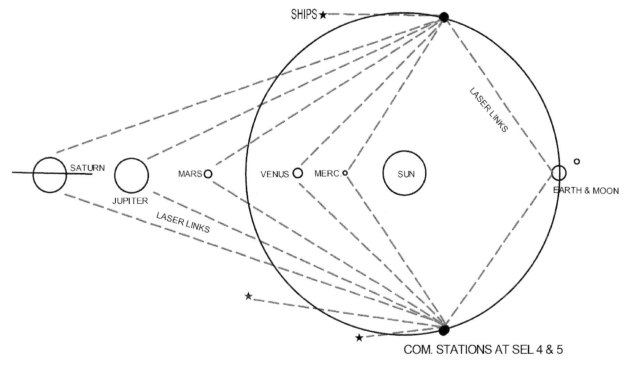

COMMUNICATION LINKS MAINTAINED BETWEEN SHIPS AND COM STATIONS
DIAL 1 + 2 DIGIT PLANET CODE + NATION CODE + AREA CODE + 7 DIGIT NUMBER
DIAL 1 + MULTIDIGIT SHIP CODE + 10 DIGIT NUMBER

Fig. 28 Laser links connect Earth, spaceships and other planets of the solar system via ground stations and orbital stations.

While there may be dedicated laser banks and optical receivers tracking the planets for communication, would there be a dedicated laser for tracking each and every single ship in space? There may be so much traffic in space someday it would seem thousands of lasers at every planet and settlement for each ship would be necessary.

Perhaps computers will store digitized messages (voice, text, email, fax) for a short time, say an hour, and then every hour send a high speed data burst lasting a fraction of a second to a ship where its computers store the data and route it to passengers. The laser could then point to another ship and transmit the previous hour's messages. This way, one laser that could be pointed in any direction could send messages to hundreds of ships. There would be some delay but there is already a delay due to light speed, so this would just have to be accepted. Time and/or code division multiplexing will be used on each laser channel.

Things might work faster. Every ten or twenty minutes the que of messages in the computer's memory could be transmitted in just a fraction of a second at hundreds or thousands of megabits per second. The high power laser or bank of lasers might be rather massive. The beam(s) could be directed at various ships with a lighter weight mirror that's easier to mechanically move around. If the laser transmits messages to 100 ships every ten minutes then it will point at each ship for six seconds. That should be plenty of time for the mirror to move around.

A high quality Skype video call uses about 500 kilobits per second.[83] The average email without attachments is about 20 kilobytes long. With standard attachments about 300 KB.[84] Many emails are sent with attachments in the megabyte range. Space telecommunication services might charge extra for those big attachments so it might be best to stick with plain text messages. A video message lasting one minute might use 30 megabits. With speeds of hundreds or thousands of megabits per second it should be possible for stored messages to be forwarded to a ship in less than a second and maybe one or two seconds at most if some huge amounts of data have to be delivered. There will be plenty of time for aiming lasers at multiple ships. With multiple laser channels or wavelengths even more data can be moved.

Space luxury liners might someday carry thousands of passengers so there should be plenty of messages flowing from ships to ships and planets to ships and vice versa. With millions of people living on various worlds there will be huge numbers of messages being sent between worlds. Since one laser beam couldn't carry all the data tunable lasers and numerous beams will be needed. Lasers are not actually monochromatic. They emit a narrow band of wavelengths. These can be separated with prisms and diffraction gratings so several lasers can each be tuned to a precise wavelength. There are numerous kinds of lasers each emitting its own narrow band of wavelengths. Most communication lasers use intensity modulation. One pulse intensity could represent "1" and a less intense pulse could represent "0" for binary data transmission. It is also possible for lasers to emit sequences of pulses with one fast set of pulses representing "1" and a slower set "0."

Receivers will use telescopes with optical set ups to separate the various wavelengths or channels and shine them on photosensitive chips that generate electrical signals to be fed to computers. A ship's telescope could stay aimed at a space station in planetary or solar orbit. Space station scopes will have to point at ships at predetermined times to receive messages. Computers that control the lasers and receivers will have to know the location of ships in space at any given time and they must be smart enough to aim ahead of the ship when transmitting, depending on the light distance, and behind the ship when receiving. The nitty gritty details of the hardware and software for all this technology will be staggering.

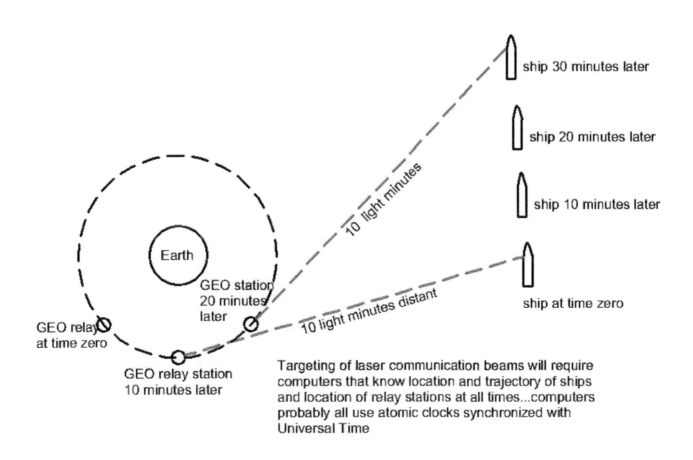

Fig. 29 Tracking ships

It seems there will be times when space liners are just a few miles or maybe a few thousand miles apart which isn't a lot on astronomical scales. I suspect that passenger liners will travel in convoys, so they really should have ship to ship radio. This capacity might be very important in emergencies, presuming the ships could maneuver and rendezvous with each other in interplanetary space. They will travel in convoys even with nuclear propulsion that can get them to Mars in 6 or 7 weeks because launch windows might be years apart and might be only a few days long. If a ship had to send messages by laser to another ship via relay stations in orbit around Earth or Mars or at an Earth-Sun Lagrange point that would really delay messages. Ship to ship radio will make real time communications between ship crews and passengers possible.

Over distances of a few miles ships could communicate with short waves, FM, VHF or UHF with simple omni-directional monopole or dipole antennas. Over hundreds or thousands of miles they might send a microwave signal with a big dish antenna that can be picked up by omni-directional monopole or dipole antennas. Once they got the other ship's attention and sent location and channel data the two ships could point their big main dish antennas at each other and send voice and video messages back and forth in real time. Multiplexed digital radio will make it possible for several conversations or data streams to be exchanged at the same time on a single microwave channel.

The interplanetary telephone networks should also make interplanetary internet possible, or vice versa. Server farms on various planets and large free space settlements could be updated regularly. Even so, there will be delays due to light speed. The internet on Mars could be a mirror of the internet on Earth. Every so many minutes the computers could transmit new images of the internet or just recent additions to the internet on each planet to the other one. They might always lag each other by 3 to 22 minutes but that's just the way it will have to be until if and when faster than light communication is achieved. If someone posts a message on some form of social media a person on a distant world will just have to wait a half hour. Given the fact that even today many of us have become accustomed to sending emails and waiting hours or days for a reply at the convenience of the receiver, it should not be very hard for people on different worlds to wait for replies. If someone places an order with an online retailer on a different world it might take just minutes for the order to get there, but weeks or months to receive the package via interplanetary freight. Human patience is not so limited that the light speed delay between worlds will cause any major anguish. As long as the computers of video providers on different worlds are kept up to date it should be possible to get the latest Hollywood movies in a reasonable amount of time for people who aren't as far out as the Oort Cloud.

Connecting from a ship might present problems. We are used to browsing the web and getting information within seconds. If one is on a ship and connects with the web on a distant world they may have to wait awhile for a response to their search engine query. That's just something space travelers will have to learn to live with. Computers on board ships might contain vast amounts of technical data needed to operate and repair the ship as well as encyclopedias and regularly updated music and video entertainment including games that passengers can access with ship wide wi-fi, so life shouldn't be unbearable for the internet and smart phone addicts of the future.

Chapter 13: Bootstrapping in Orbit

A Moon mining base, mass driver lunar launcher, mass catchers that can capture and transport raw material through space and constellations of communication and navigation satellites serving these won't be worth much if there is no infrastructure for building Space Solar Power Satellites in GEO. There will have to be manned stations consisting of inflatable modules and unpressurized robotic stations. There will have to be regolith refineries where metals, glass and ceramics are produced. Autonomous AI robots and teleoperated robots controlled by crews in space and on the ground will be necessary. To build 100 SSPSs in 25 years there would have to be at least four refining, manufacturing and construction complexes in GEO with each one producing one powersat per year. This could cost a lot of money.

The way to cut costs would be to orbit only the minimum required mass of equipment and use lunar materials to build up the infrastructure. Bootstrapping in space can be done just as it can on the Moon once raw materials are available. If 1000 tons of cargo is launched to LEO with Falcon Heavy rockets the price tag will be $1.7 billion. Orbiting another 300 tons of ion tugs and reaction mass would cost $500 million. It might be possible to reuse ion tugs that hauled cargo to the Moon to cut costs. It will only be necessary to propel the cargo from LEO to GEO. High thrust lander rockets and propellant for them won't be necessary.

Bootstrapping in GEO will be a lot like bootstrapping on the lunar surface. Raw regolith will be transformed into oxygen and metals with Supersonic Dust Roasters and All Isotope Separators. Rotating space stations spinning fast enough to produce perhaps lunar equivalent gravity will make it possible to pour molten metals and basalt into molds. The SDR requires gravity to work but there is a microgravity version of the AIS. Gravity also helps things stay put so there is no problem with things floating away when slightly disturbed. Dealing with coolants and lubricants when machining parts will also be easier with some "artificial gravity." There will be Coriolis effects to compensate for when pouring molten materials, but that's not a showstopper.

Cryogenic equipment for liquefying oxygen and things like rolling mills, drop forges and extruders will be manufactured in space by casting and machining steel parts. Powdered metals will be formed into all sorts of parts with 3D printers. Basalt will be sintered and cast in iron molds made by 3D printing. It will also be drawn into fibers. Robots and human workers will assemble things.

The rotating space work-stations could be made of aluminum I-beams fastened together with steel rivets or bolts. The beams could be made by rolling aluminum slabs produced in space. The first aluminum beams and fasteners would be rocketed up to GEO and the first station could be constructed with payloads from Earth. After materials are produced and heavy equipment like rolling mills are made at the first station by casting and precision machining as was done on the Moon, more I-beams could be made. Steel rivets or bolts could be produced in space and the beams could be assembled by using mechanical arms, robots and space walking humans. The new stations could be propelled to other locations in GEO with solar electric ion propelled tugs. Solar energy is available constantly in GEO so work-stations and ion tugs will have a bounty of energy available at all times. This could make work more efficient in GEO than in LEO or on the Moon where sunshine is blocked half the time.

Rotating Space Manufacturing and Construction Station for Processes that Benefit from "Artificial Gravity" and Pressure as well as Vacuum Conditions

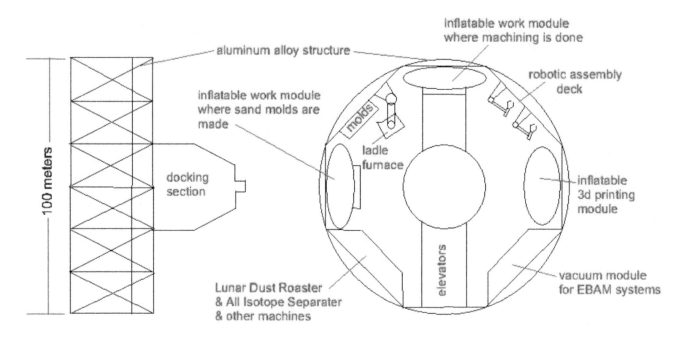

Fig. 30 Station is about 100m diameter and rotates at 1.7 RPM to produce lunar equivalent "gravity."

Manned and teleoperated orbital maneuvering vehicles (OMVs) with mechanical arms will be used to assemble powersats. Non-rotating weightless work platforms consisting of beams riveted or bolted together could be used for processes that don't require gravity. Space suited workers will use manned maneuvering units (MMUs) to get around in space and use specialized tools to perform some tasks.

Aluminum, titanium and magnesium will be used for structures like the space frames that support the millions of square meters of silicon solar panels that compose a powersat. Pure aluminum will also be used for doping p-type solar panel material, electrical wires and cables. Aluminum alloyed with magnesium, silicon and/or manganese will also be used for structures. Rolling mills will churn out the aluminum alloy I-beams and tubes that make up solar power satellite frames. Titanium could be used for fasteners and connections at high stress points in the structure. Reflective magnesium foils or sheets might be made to concentrate sunlight onto solar panels. This could raise solar panel temperatures and reduce efficiency. Cooling systems would become necessary and this might add too much complexity and cost to the powersats.

Fiber reinforced basalt composites might be created and these might be used for structures. Basalt parts could be cast and it may be possible to roll or extrude red hot softened basalt. Basalt and glass can make electrical insulators for power wiring and cable systems. These materials will not involve by-product oxygen.

Since regolith is about 40% oxygen by mass, smelting metals will result in more oxygen than is necessary for life support. That oxygen could be compressed and stored in high pressure cylinders for cold gas thrusters on manned maneuvering units, OMVs and construction robots. Oxygen could also be sold to lunar tourism companies for their rockets. The regolith is about 21% silicon and this might prove to be more than is needed for solar panels and SSPSs. With some liquid hydrogen "piggybacked" into orbit with passengers, excess silicon could be converted to silane. Regolith has plenty of iron, about 12%, and calcium, about 8%. With the exceptions of calcium wires and steel fasteners and rocket motor casings there won't be much use for these two metals. Iron and calcium will burn in pure oxygen. They could serve as low performance rocket fuels when mixed with silane carrier fluid. The powersat builders could make some extra money by selling oxygen and fuel to the lunar tourism companies and perhaps Mars bound settlers.

A synergy between SSPS building, orbital habitat construction, lunar tourism and the settlement of Mars and other places in the solar system beyond Earth-Moon space exists. Not only propellant can be produced but materials for building ships and space stations can be supplied. Eventually there won't be a need for more powersats and the Moon miners and space construction companies will need a new market to sell their services and products. Orbital habitat for multi-million dollar condos and time shares could be big business. Since space stations don't have to go anywhere mass is less critical than it is for a spaceship. Iron and steel, glass, basalt and cement/concrete can be used for orbital habitat building. Reboost won't be a problem for these massive constructs if highly efficient electric propulsion systems like ion drives or VASIMR are used to overcome the small amount of atmospheric drag at altitudes of several hundred miles. Vaporized magnesium might be used as reaction mass.

Asteroid mining becomes more lucrative with a substantial amount of industry and infrastructure in space. The main asteroid products-water and hydrocarbons, could be sold to space dwellers in need of drinking water, chemicals and plastics as well as chemical rocket propellants and hydrogen for nuclear-electric propulsion (NEP) systems for ships to Mars and beyond. Since electric propulsion generates low thrusts, chemically propelled boosters may be desired to get ships up to escape velocity. The electric drive would operate continuously for weeks until the ship reaches top speed. Nuclear fuels could come from the Moon instead of being launched from Earth. Aluminum, magnesium and titanium from the Moon, once used for powersats, could be used to build large spaceships and orbital habitat. Space luxury liners and orbital resorts will emerge. Space construction techniques and technology will be developed to a high art by powersat builders.

Bootstrapping in geosynchronus equatorial orbit (GEO) will be a lot like bootstrapping on the Moon but there won't be any need for robots that can microwave regolith, mine for iron fines and volatiles, and bulldoze or excavate huge amounts of regolith. There won't be any wheeled vehicle and railroad building or mass driver launcher construction. Bootstrapping in orbit might be simpler than bootstrapping on the Moon. Everything will focus on work-station replication and solar power satellite construction. There will be many robots capable of maneuvering around in space, OMVs and MMUs. Solar panel mass production and the machines needed will be challenges faced by bootstrappers on the Moon and in GEO. Without a doubt, space industry will have to develop cost effective ways of mass producing solar panels on the Moon and in space or there will be no progress on the high frontier. Substantial amounts of phosphorus will be needed for solar panel n-type material. Regolith that contains KREEP minerals is richer in phosphorus than typical regolith. Moon miners might dig for KREEP laden material and extract phosphorus and send it to GEO stations by way of special rocket delivery.

Since teleoperation in GEO only involves a fraction of a second delay time instead of the nearly three second delay time for Earth-Moon teleoperation, it will be possible to do most work in space by remote control via ground stations. Skeleton crews to fix robots or do fine hand tasks will be all that are needed in space. Humans are expensive to transport and maintain so putting only the minimum number in space will keep down costs. Radiation could be a problem since GEO is not protected by Earth's magnetic field as is LEO. Orbital habitat must be surrounded by thick layers of regolith or water. Weightlessness will also have deleterious effects. Human stays in space might be limited to a few months. This is one of the many reasons robots are cheaper. They can go one way and not incur return transportation costs. They can remain in space until they no longer work. They don't need to sleep. They don't need pressurized habitat, radiation shielding, food or water. When they die they can be recycled. Some occasional maintenance by other robots or humans is all that's needed. Major overhauls involving the replacement of joints, bearings, electric motors and electronics might make it possible to use robots for decades in the rust and corrosion free vacuum of space. Finally, robots won't demand hazard pay or insurance benefits. Smart machines will be expensive, but the investment should be worthwhile.

It's impossible to predict what AI will be capable of in coming decades. Humans might not be needed in space at all. If the need for human intervention arises it might be possible to do the job in space with teleoperated systems. Three equatorial radio stations 120 degrees of longitude apart connected to ground control stations via the internet could make it possible for humans to effect remote control at any time anywhere in GEO. The use of robots may not be as exciting as manned activities in space, but it could turn out to be much more practical and cost effective. While the nearly three second delay time for operating robots on the Moon from Earth might make humans essential on the Moon, this might not be the case in GEO and LEO. However; if AI soon rivals the human brain human controllers on the ground might just transmit orders to the AI robots on the Moon and let the machines figure out how to do the job. Mars is too far away for teleoperation, so most space prognosticators envision humans on Mars. Will AI obviate the need for humans or at least reduce the number of humans required on Mars? This scenario is not as exciting as human space settlement, but it could prevent a lot of injury and death. For people willing to take chances, space travel is really tempting. Robots could safely pave the way for millions of space tourists someday.

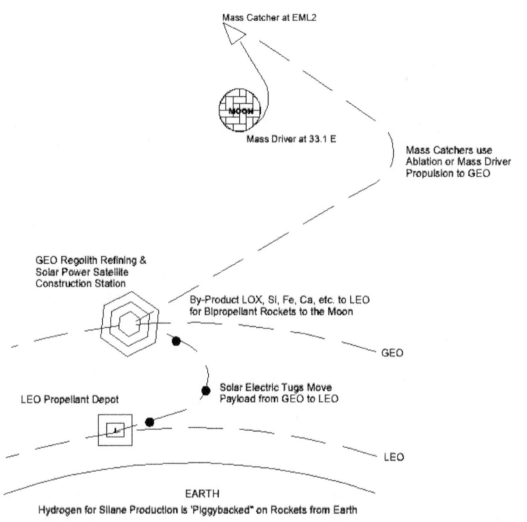

Fig. 31 Transportation of materials from the Moon to GEO and LEO.

The mass catchers that move regolith loads to GEO and the tugs that move oxygen and metals down to LEO will be robotic. These vehicles will make long slow trips between places in orbit and the tugs from GEO to LEO will spend a lot of time in the Van Allen Belts. This is not desirable for manned spacecraft. Aluminum, titanium and magnesium along with smaller amounts of iron for steel and silicon for solar panels will be used for powersat construction in GEO. The seemingly useless by-product oxygen, silicon, iron and calcium will be sold to companies operating rockets in space. The cost to the powersat builders of storing and disposing of all these wastes will be eliminated. The rocket transport companies will take this stuff off their hands for free or even pay a small price for it. Construction companies in LEO might want some of those materials also.

Chapter 14: Recycling

Critics of space industrialization and settlement have exclaimed," We will trash outer space the way we have trashed the Earth." This is unlikely. The Moon, outer space and other worlds offer such austere environments that waste cannot be tolerated. Space dwellers will have to recycle everything or they might not survive. Energy efficiency will be called for too. Insulation and the use of waste heat from one industrial process to power another one would be useful. Although solar energy is plentiful and free, within the orbit of Mars at least, the technology to harvest it is not. Solar panels will be mass produced and the materials from old panels will be recycled. Uranium and thorium for nuclear energy exist on the Moon in KREEP terranes but they are only present in parts per million. Enormous amounts of regolith will have to be excavated and processed with SDR-AIS devices to get decent quantities of nuclear fuel. Breeder reactors will be needed to convert most of the uranium in the form of fertile U238 to fissile plutonium. Thorium 232 will have to be converted to fissile U233 in breeder reactors. It will be necessary to reprocess and recycle nuclear waste to get fissile elements and use them to generate more energy. There are nuclear reactors than can run on nuclear waste and they will be applied also. Very little nuclear waste will result. There is no life, air or water to poison on the Moon so nuclear energy could someday be utilized there. Nuclear fuels will also be needed to power ships that travel all over the solar system. It is not known if there is life on Mars but there is a thin atmosphere and sub-surface permafrost that we might want to prevent contamination of by radioactive waste. Free space settlements will use solar energy.

Discarded metal items will be melted down in solar and electric furnaces to be cast again or formed by various processes like rolling and forging into new products. Steel and aluminum are recycled on the resource rich Earth today. It is cheaper to recycle them than produce fresh metal from ore. This will certainly be the case on the Moon. Things made of alloys containing expensive imports like copper or vanadium will be too valuable to just throw away. There might be small furnaces for melting just a few pounds of metal from a worn out part made of an exotic alloy and furnaces for melting huge tonnages of more common metals. When vehicles made of steel, titanium and aluminum alloys get old they will be dismantled and all the separate metal parts depending on their composition will be sorted out and melted down unless the parts are still good and can be used again. There won't be junkyards full of rusting scrap on the Moon but there might be warehouses stocked with second hand parts.

Paper will be a valuable commodity. Paper waste will be shredded, mixed with water to make pulp, bleached and laid out in sheets that are then baked to make new paper for newspapers and documents. Given the value of paper, most written communication will be done electronically with smart phones, tablets, laptops and Kindle readers. Artists will need something besides paper to paint on. Peter Kokh has experimented successfully with sodium silicate based paint on glass. He has proposed several different colorizing agents that could be produced on the Moon.[85] Cast basalt tablets and metal plates might also be painted on. Carvings on stone and engravings on metal might be common.

Glass will be recycled. Jars and bottles for foods and drinks as well as drinking glasses and other dinnerware will be washed, sterilized and reused. Broken glass discards will be washed, ground up, melted down again and recast to make new items. This might require energy but it won't mean excavating regolith and extracting silicon dioxide, calcium and sodium again to make fresh glass, so it should be more efficient to recycle broken glass.

Plastics are lightweight, durable and easily shaped by injection molding. When there are no substitutes plastic will be in demand. Plastic is cheap on Earth but in space it will be pricy due to the scarcity of light elements. Plastic bottles, drinking vessels, pitchers and bowls will be washed and reused. Thermoplastics will be sorted, washed, melted and injected into metal dies to make new things. Biodegradable Poly-Lactic Acid (PLA) items will be chopped up and composted. Plastics that cannot be recycled will either simply not be used or they will be decomposed in extremely hot plasma arc furnaces to break them down into hydrogen, nitrogen, chlorine and carbon. Nitrogen will be used to make air and fertilizer. Chlorine will be used for chemicals, silicone production and synthetics like PVC. Hydrogen and carbon could be burned with oxygen to get water and CO_2 that can be used for farming or making new chemicals. Synthetic fibers will be recycled by ripping up old clothes and spinning new thread. Natural fabrics will also be recycled in this way. As for fibers that can't be recycled they will be composted if possible or destroyed in the plasma arc furnace and broken down into their constituent elements that will then be used to make new things.

Air and water will be recycled. At first, physio-mechanical systems that involve lithium hydroxide cannisters to absorb CO_2 from the air and/or cryogenic systems to freeze CO_2 and other impurities out of the air will be used. Oxygen will be extracted from regolith. Grey water from kitchen sinks, showers, hand basins and tubs will be screened, filtered and piped to toilets for flushing. These systems can reduce water use by 30% to 50%. The black water from toilets will be sand filtered, distilled and sterilized with UV light, heat and pressure. Water from dehumidifiers could be treated the same way to prevent the spread of airborne disease. Imported food that is dehydrated or freeze-dried will be rehydrated with the same water over and over again. For the sake of morale there will be some canned and frozen whole foods. Wheat flour, corn meal and oats can keep for a long time, so there might be some fresh baked breads, biscuits, muffins, hot cereal and cookies. Pasta and bottled or canned spaghetti sauce keep for a long time also. Let's not forget dried fruit, nuts, dried beans, powdered milk and eggs. In time, Closed Ecological Life Support Systems (CELSS) will be created. Air will be purified by green plants that remove volatile organic compounds from the air and use photosynthesis to convert CO_2 and water to oxygen and food. Fresh food will be produced. Workers and eventually their families and tourists will be much happier.

Grey water systems will still be employed for the sake of efficiency. On average in the US, direct indoor water use (water from the tap, toilet, dishwasher, etc.) adds up to about 138 gallons per household per day, or 60 gallons per person per day. This is based on a two person household. They would use about 120 gallons per day but another 17 gallons is lost due to leaks. The most common sources of leaks are toilet flapper valves.[86]

Appliance/Device	Household per Day
Toilets	33 gallons
Showers	28 gal.
Faucet	26 gal.
Washing machine	23 gal.
Leaks	17 gal.
Bath	4 gal.
Dishwasher	2 gal.
Other	5 gal.
Total	**138 gallons**

Leaks must be stopped in a Moon base. That would be too wasteful, but they will probably happen anyway. Workers will replace old flapper valves, fix dripping faucets and stop leaks from pipes. The 5 gallons for "other" is water for evaporative cooling, dehumidifiers and hot tubs. It doesn't seem likely that there will be hot tubs in early Moon bases but there might be dehumidifiers and evaporative coolers. Showers, faucets (to get water for drinking, cooking, washing, brushing teeth, etc.), baths and dishwashers use 28 gallons, 26 gal., 4 gal. and 2 gal. respectively for a total of 60 gallons a day for two persons. The same amount of grey water will be produced. Since toilets use 33 gallons that leaves 27 gallons of grey water for other things. Grey water could be treated in settling tanks to remove dirt, filtered and sterilized. This water could be used in the laundry. According to the above list the laundry will use 23 gallons leaving 4 gal. of grey water for plants and things like mopping and steam cleaning. New Energy Star certified front loading washing machines can use as little as 13 gallons per load. If the laundry only uses 13 gallons per day for two people that leaves 14 gallons of grey water for non-food plants like bamboo, hemp and cotton and for mopping and steam cleaning of floors and sometimes walls.

It makes sense to use filtered, sterilized (heat, pressure and UV light) grey water for laundry. The laundry discharge would go to the blackwater tank and then to the yeast and bacteria bioreactor where microbes eat up the soap and other substances in the water. Water from kitchen sinks will contain organic matter like grease and bits of food. Hand basins and showers will add hair, skin cells and soap to the grey water. These can be settled out, filtered, and the residue can be sent to the bioreactors when the filter is changed or the sludge is drained from the settling tanks.

Water cannot be sterilized with chlorine, ozone or any other disinfectant chemicals because they would kill the yeast and bacteria in the bioreactor that produces hydroponic nutrient solution for crops and algae. Liquid soap might be used in the washing machines but bleach and chemicals should not be used. Heat, pressure and UV light are the only ways to safely sterilize water in a Moon base.

Floors and other surfaces would be cleaned with steam instead of bleach, ammonia or other chemical disinfectants so that the interior air does not become poisonous and the water polluted. Hydrogen peroxide solution might be used. That only breaks down into oxygen and water. Grey water will work for steam cleaning. Fresh water consumption can be reduced with low flow shower heads. This means less grey water but grey water consumption can be reduced with low flow toilets. Evaporation from showers, sinks and steam cleaning, etc. will occur and that water would be condensed from the air, sterilized and sent to the fresh water tanks. All in all, we can expect grey water systems to cut fresh water usage by about half.

What about that 5 gallons of "other" water and 17 gallons from leaks? That's 22 gallons of water to account for. Where does that water go? It seems likely most of it will evaporate and be recaptured in the condensors. If water leaks from the flapper valves in toilets that could drain the grey water storage tanks and automatic devices would fill them back up with fresh water thereby defeating the purpose of a grey water system. The entire plumbing system will require serious maintenance to prevent leaks. Leaking pipes will leave water that evaporates and some of that water will be mopped up and dumped down the drain into the grey water system. Dripping faucets will let fresh water trickle down into the grey water tanks which could overflow into the blackwater tanks thereby wasting fresh and grey water. As for the "other" water, grey water might be used for things like dehumidifiers.

Black water will go to yeast and bacteria filled bioreactors where the waste is decomposed. The effluent from the bioreactors will go to algae tanks that convert CO_2 into oxygen and provide feed for fish and prawns.[87] Water will be processed further to make it drinkable, but it will not be sterilized with chlorine and ozone. These would make the water toxic to the yeast and bacteria in the bioreactors and could even poison the fish and prawns. Potable water will be sterilized with UV, heat and pressure.

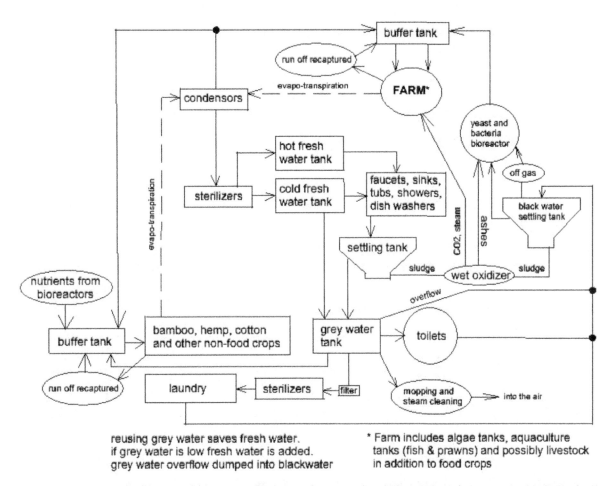

reusing grey water saves fresh water.
if grey water is low fresh water is added.
grey water overflow dumped into blackwater

* Farm includes algae tanks, aquaculture tanks (fish & prawns) and possibly livestock in addition to food crops

water from condensors, run off, grey water and nutrients from bioreactors are mixed in buffer tanks to make hydroponic solution applied to roots and piped to algae tanks

Fig. 32 Water recycling system

Salt is an item that will be recycled. In ancient times salt was so valuable Roman soldiers were paid with it; hence the expression "worth your salt." It can be used for things from flavoring and preserving food to cleaning coffee pots, making soap and mixing up saltwater for gargling, brushing teeth or cleaning out nasal passages with a Netti pot. It is also necessary for dear life itself. Sodium is an essential human nutrient. There is sodium on the Moon that could be combined with imported chlorine to make salt. If sodium is constantly mined and reacted with chlorine to make new salt that is excreted into the water loop of the CELSS it will build up in the water loop and this could be toxic to the living things in the bioreactors, algae tanks and aquaculture tanks. Since we don't want salt poisoning in the loop it will be necessary to control salinity with reverse osmosis pumps that use filters made from bamboo perhaps to remove salt from the water. Halophytes, salt loving plants, might also be cultivated to remove salt. Since excess salt will be extracted from the water loop it could be sterilized and used over and over again. Mining for sodium and importing chlorine would no longer be needed. Salt will come directly from the RO pumps and it could be extracted from plants by charring them and washing the ashes with water to recover salt. Sometimes this yields sodium oxide, washing soda. Other times sodium hydroxide, lye, is obtained. Sodium in the form of salt, washing soda and lye used for soap can cycle endlessly.

Lunar ground vehicles and spacecraft will not rust in the vacuum. Besides the lack of air and moisture they will be made of corrosion resistant aluminum and titanium alloys primarily. Building them will not be cheap. These conveyances will be repaired rather than replaced and undergo periodic overhauls and upgrades. Ground and space vehicles should last for decades. The worst wear and tear might occur in wheel bearings and mechanical joints of ground vehicles and excavators due to the abrasiveness of lunar dust. Silicone lubricating grease and plastic seals around bearings and joints could protect them. Batteries will have limited lifetimes; even solid state lithium ion batteries. Old batteries will be dismantled and their constituent elements recycled. Vehicles that are so old that they cannot be overhauled will be taken apart bolt by bolt and even cut up with torches or lasers, their parts separated by material types, and finally melted down and formed into new products.

Old electronics will be disassembled and processed to get their precious metals. This is done profitably on Earth today. On the Moon and in space recycling electronic scrap will be even more important. Eventually, metallic asteroids will be mined for precious metals, yet they will remain too valuable to waste.

Chapter 15: Nuclear Power

The Basics

Climate change could lead to a rebirth of the nuclear power industry since nuclear power plants don't emit greenhouse gases. Modern light water reactors in strong containment buildings will not result in a Chernobyl scale accident. There are also inherently safe molten salt reactors (MSR) that cannot melt down. The nuclear fuel in these MSRs is dissolved in molten salt. If the reactor overheats due to a cooling system failure the salt and fuel expands, neutron flux density decreases, and the fission process slows down. In the worst case, the hot salt could melt a plug and the salt and fuel would flow out of the reactor and into heavy containers without escaping into the environment.

Most people don't understand the functioning of a reactor. Neutrons that cause fission of the atomic nucleus are not like cue balls that smash into racked up billiard balls when the pool shooter breaks. Neutrons are slowed down by a moderator made of carbon, beryllium oxide, zirconium hydride, water or heavy water. These slow neutrons are absorbed by the uranium or plutonium nucleus. The nucleus then forms an unstable configuration and decays rapidly; splitting up into lighter elements and releasing energy as well as more neutrons to cause more fission reactions.

In typical light water reactors, the coolant water is also the moderator. If the water is lost the fission reactions slow down or stop. However, the radioactive by-products continue to decay and release heat and the reactor melts down. The molten core might not ever escape from the thick steel walled pressure vessel that contains it and if it does it won't get out of the reinforced concrete containment structure barring a freak accident like an earthquake, meteor impact or heavy airplane crash.

Nuclear power plants are actually very safe, despite the publicly perceived dangers. Radioactive waste is a problem that has not been dealt with for political reasons for too long now. The waste can be vitrified, that is mixed into inert glass. Then it can be placed in 60 ton steel casks that will be placed in geological repositories. In one test, a locomotive was slammed into a cask at 70 mph and the locomotive was demolished while the cask was not damaged. Thus, transportation of nuclear waste can be done safely.[88]

Nuclear waste contains unreacted or "un-burnt" uranium 235 and some plutonium that formed when U238 absorbed neutrons. There are other fission products that have half lives of millions of years and others that have much shorter half-lives. If nuclear waste is reprocessed to extract the unburnt U235, plutonium and other long half-life elements, those elements can be formed into fuel rods and put back into the reactor to generate energy. The elements with short lifetimes could be stored in stable underground rock structures and they will decay until they are no more radioactive than natural ore in less than a thousand years.[89] The only problems are that in the USA no nuclear waste repositories have been licensed and the cost of reprocessing is greater than the cost of using virgin uranium, so it isn't done.

Nuclear Power on the Moon

On the Moon there is no life and no air or water to contaminate. If nuclear waste did escape into the environment it would not dissolve into ground water and spread because there is no ground water on the Moon. Permafrost has been detected all over Mars and we must wonder about nuclear waste giving off heat and melting the frost and spreading. Nuclear power could be used on the Moon safely. The environment there is already so highly radioactive due to cosmic rays that some uranium would just be a drop of water in a swimming pool. Our Earth's geomagnetic field deflects many cosmic rays and the thick atmosphere blocks the rest of them from reaching the surface with the exception of some rare ultra-high energy rays. Inhabitants of the Moon will need to cover their shelters with about nine tons of regolith per square meter or a layer about six meters (20 feet) thick. They might also locate bases inside lava tubes.
Since the Moon is drenched with solar energy half the time at lower latitudes, never obscured by clouds or dust storms, and almost 90% of the time in polar regions, the need for nuclear power on the Moon might not ever exist. Solar power plants could emerge all over the Moon someday to feed a globe girdling power grid that supplies power 24/7 all year round. Small nuclear power plants might still be built at remote outposts and mine sites that can't tap into the power grid.

A small nuclear power plant could be really helpful in the early days of lunar bootstrapping. It could supply waste heat to keep equipment warm during the two week long super cold temperatures of the lunar night as well as electricity. Since the present political situation would probably make it impossible to rocket re-entry safe vaults full of nuclear fuel into space, it seems that early lunar industrialization will depend on power storage in the form of batteries, fuel cells and/or flywheels as well as thermal storage in blocks of cast basalt, concrete or molten salts to get through the lunar night.

If nuclear fuels can be produced on the Moon the dangers of rocketing them up from Earth can be averted entirely. Nuclear power will be essential for the electric propulsion systems, be they ion drives or electrothermal VASIMR drives, of future spaceships that travel to Mars and beyond. There is a kind of material on the Moon called KREEP that contains uranium and thorium. KREEP is an acronym. K = potassium REE = rare earth elements and P = phosphorus. This material is enriched in these elements compared to typical regolith and it is most common at the edges of Oceanus Procellarum and Mare Imbrium. Potassium and phosphorus are needed for agriculture. Rare earth elements have many industrial uses including neodymium and samarium-cobalt magnets for EV motors and Amplitrons. There is not a lot of uranium and thorium in this stuff; just about 4 ppm and 10 ppm respectively. Thus, if a million tons of KREEP terrane is excavated about 10 tons of thorium will be gained and 4 tons of uranium. Natural uranium in this solar system consists of about 0.7% U235 and 99.3% U238. That would result in 28 kg of U235 and 3,972 kg of U238. The U235 can undergo fission and serve as a nuclear fuel. It is called fissile. The U238 is not fissile but it can be converted to plutonium which is fissile. It is called fertile. Thorium is not fissile either, but it is fertile. It can be converted in a breeder reactor to fissile U233.

Typical modern light water reactors use low enriched uranium with about 4% U235. With 28 kg. of U235 it would be possible to make 700 kg. of low enriched nuclear fuel. That's not much compared to the 100 tons of low enriched uranium loaded into a reactor at a 1000 MWe (megawatt electrical) power plant. Will lunar nuclear fuel production be worthwhile? It could be if the potassium, phosphorus and rare earth elements can also be sold to lunar settlers, entrepreneurs and visitors for various purposes. A lot of oxygen, silicon, iron, calcium, aluminum and magnesium would also result when that KREEP material is refined and those too could be sold.

This isn't the end of the story for nuclear fuel. A light water reactor will leave nuclear waste with U235 in it and some plutonium if it is not reprocessed. When the U235 content drops after about a year of operation the reactor can't generate enough power to make the plant worthwhile because of the accumulation of neutron absorbing waste products. So it is shut down and refueled. A fourth of the fuel rods, 25 tons of them, are replaced. The spent fuel rods go to waste, but that's cheaper than reprocessing.

On the Moon it seems reprocessing would be essential. Even better, breeder reactors could be put to work. Breeder reactors use a core immersed in a coolant pool of molten sodium. The sodium doesn't moderate the neutrons as well as water or carbon can, but the high speed neutrons are absorbed by U238 in a "blanket" around the core and this fertile uranium isotope becomes fissile plutonium. The core contains U235 and eventually it could be loaded with plutonium to make neutrons for breeding more nuclear fuel from U238 and generating power. Since the fast neutrons in a liquid sodium cooled breeder are not captured as easily or as often by U235 nuclei, a higher percentage of U235 on the order of 10% to 12% is used in breeder reactor fuel rods. If a core of U235 or plutonium is surrounded by a blanket of fertile thorium it becomes possible to breed fissile U233. A mixture of plutonium and U233 could be placed in the core and plenty of thorium in the blanket could be converted to nuclear fuel. With each breeding cycle more nuclear fuel will be produced than is consumed. Molten salt breeder reactors are also possible. Another interesting possibility is the use of electro-nuclear breeding. Solar energy would be used to power a particle accelerator in the free vacuum of the Moon that blasts heavy metal targets with deuterium nuclei. When the deutrons impact heavy metal nuclei neutrons are released by the process of spallation. The neutrons could then crash into thorium targets and convert it to U233. Whether or not this would be economical is unknown.

The Moon miners could go into business selling nuclear fuels to interplanetary spaceship owners. Nuclear electric propulsion (NEP) would use far less reaction mass than chemical rockets and it could deliver higher speeds to shorten transit times to other planets. While solar electric propulsion (SEP) might be considered for ships in the inner solar system, it must be remembered that solar energy is only 43% as intense at the distance of Mars from the Sun and it is only about 4% at Jupiter's distance and a bit more than 1% at Saturn's distance. NEP will be essential for travel into the outer solar system. Time must also be considered. A minimum energy orbit from Earth to Jupiter could be attained by chemical rockets but will take 2.7 years and for Saturn 6 years. Radiation exposure, psychological stress, life support and supply problems would make this impractical for human travelers. A six month journey to Mars could be shortened to 39 days with fast NEP using VASIMR and vapor core reactors generating 200 megawatts of electrical power.[90] Voyages to Jupiter and Saturn or their moons more properly could be completed in less than a year with powerful enough reactors and lunar nuclear fuels.

Moon miners could always dig up more than a million tons of KREEP annually too. The reactors used to breed fuel for sale to spaceship owners could also generate power for the reprocessing facilities, fuel rod fabrication plants and mine operations. At least the Moon miners at the nuclear communities won't have to buy electricity from anyone.

Also, we must wonder about the presence of uranium and thorium in asteroids. These heavy metals tend to sink to the core of planets. The asteroid Psyche may be the exposed core of a shattered small planet. Many M-type or metallic asteroids might be rich in uranium and thorium as well as platinum group metals. If not, then perhaps mining metallic asteroids for nickel, cobalt and platinum group metals will yield uranium and thorium as by-products of great value. The planet Mercury is very dense and probably consists largely of heavy metals so uranium and thorium might be found there. Finally, most uranium and thorium ores on Earth are formed hydrological processes. Since there was once water on Mars there might be uranium and thorium concentrations and ores containing other valuable elements.

Vapor Core Reactors

Science fiction authors have envisioned fusion powered spaceships. This might never be realized. First of all, we may never get economical controlled nuclear fusion power. Fifty years ago the experts said we'd have fusion in fifty years and now they are predicting it in another fifty years. Some have quipped," Fusion is the power of the future and always will be." Deuterium-tritium fusion releases floods of high energy neutrons that could damage the reactor, shorten its lifetime and result in poor economy. Helium 3 fusion does not release neutrons but it is much harder to make happen. Another problem with fusion is that magnetic containment reactors with their massive superconducting coils seem to be inherently massive devices. The International Thermonuclear Experimental Reactor will be the largest fusion reactor in the world when it is completed. It will weigh 25,000 tons and its cost, shared by 35 nations, is $23.7 billion.[91] The reactor will generate 500 megawatts thermal with a heating input of 50 MW but the total amount of energy required to drive the reactor will be several hundred megawatts.[92] It will not generate any electricity. If it could generate electricity, that 500 MWt would only produce 150 MWe to 250 MWe depending on how efficient the heat exchangers and turbines are.

Vapor core reactors (VCR) using gaseous cores of uranium tetrafluoride, UF_4, and beryllium neutron reflectors have been designed at the University of Florida's Innovative Nuclear Space Propulsion Institute that could generate more than 1 kWe per kilogram of entire system mass. That includes the reactor, magnetohydrodynamic electricity generators (MHD), plumbing, pumps, heat exchangers and waste heat radiators. Even more power might be extracted from these reactors with compact supercritical CO_2 turbines to spin generators. These turbines are only about one tenth as large and heavy as comparable steam turbines.[93] Also of interest, VCRs operate at very high temperatures and they reject waste heat at high temperatures. Since emissivity goes up to the fourth power of the absolute temperature, waste heat radiators will not have to be as big as football fields or be burdensomely massive.

A 150 MWe to 250 MWe VCR system could have a mass of 150 to 250 metric tons *or less* rather than 25,000 tons. That's enough power to propel a ship to Mars. Even if fusion power becomes a reality before the end of the 21st century, low mass VCR fission systems may be preferred for NEP ships. Such optimism must be tempered by the fact that no space VCRs have been built and demonstrated, but hope springs eternal. The solar system probably contains enough uranium and thorium for thousands of years of space voyaging. There is enough uranium in seawater to power civilization on Earth for millions of years, but the problems of waste disposal and potential nuclear accidents along with the difficulty of mining all that seawater will probably prevent that from ever becoming objective reality.

Legal Matters

The fear of nuclear power has something to do with confusion with nuclear weapons which are monstrosities. It also has something to do with the fear of cancer and the fear of genetic mutation. Nobody wants deformed offspring due to radiation exposure and a curse on all subsequent generations. Medicine might ultimately cure all cancers and genetic engineering might rid humanity of all inherited diseases and deformities without turning to eugenics programs that breed people like livestock. Radiation won't be so universally feared when this is achieved.

Even so, nobody wants a nuclear reactor to come crashing in from space. No businesspersons want to be sued because one of their nuclear vessels crashed into or burned up in the air over a large city. Ships with NEP will probably not be allowed in low Earth orbit. International law might forbid reactors within 20,000 miles of Earth's surface. In this case, travelers to Mars or the outer solar system will just have to fly up to LEO then transfer to space taxis that take them to NEP ships waiting in GEO. That will only take a couple of hours. Inter-lunar ships that travel from LEO to EML1 or LLO space stations and back will rely on chemical propulsion with propellants from the Moon and/or near-Earth asteroids.

Safety will not be sacrificed in space. The lawsuits would destroy space industry if precautions are not taken. Low budget space travel that bypasses safety regulations does not make for an appealing future. There are always greedy people who cut corners. Governing bodies and agencies that inspect and certify spaceships, stations and equipment with the force of law behind them will be called for. Anarchy in space does not make for an appealing future either. Even the old American western frontier had U.S marshals, sheriffs, deputies and territorial judges. Similar institutions will exist in space. Civilization depends on law and order. There wouldn't even be any currency if there was no government. Somehow, I don't think anarchy and bit-coin or other crypto-currencies will make for a peaceful, safe existence on the new frontier of outer space, the Moon, Mars and beyond. When we consider the possibility of terrorists stealing nuclear materials for bombs the need for law and order is even more obvious. There will be corporate security teams as well as government police forces. Bio-terrorism is another threat that might be even harder to counter than nuclear terrorism. A small test tube loaded with deadly microbes that can't be detected with radiation sensors will be easy to smuggle in. Releasing germ weapons in the closed confines of ships, space stations and surface bases will be extremely dangerous. Besides baggage checks and extreme vetting of all persons going into space, what can be done? Perhaps advanced lie detectors or brain scanners that can weed out criminals will exist in the future.

The high frontier might be a Libertarian panacea for merchants and entrepreneurs looking to make fortunes, but there will still be the need for law, order and legal tender. Digital security systems will be a must. Armed police and armed private security forces will be essential. Insurance coverage for everything from Moon quakes to terrorist attack will need to be purchased. At least nobody will have to worry about floods, lightning strikes, tornadoes and other weather events. Suicide bombers and saboteurs are not deterred by execution or being shoved out into the vacuum. Frontier justice might involve harsher penalties than death. With a little imagination one can dream up all kinds of tortures such criminals would be worthy of.

Nuclear energy cannot be forbidden because of fear and terrorism. Elaborate safety and security systems will stave off danger. Travel to the outer solar system without nuclear power will be impossible. Even in the inner solar system nuclear power for electric propulsion systems has advantages over solar electric propulsion. The NEP systems will work in the shadow of a planet or asteroid and the ship doesn't have to be oriented in any particular way to aim the solar panels at the Sun. Solar panels degrade in a few years' time while nuclear powerplants can operate for decades. Unless rich hydrothermal or sedimentary deposits of uranium rich ores are found on Mars, many millions of tons of lunar KREEP materials must be mined. It might also become possible to ship nuclear fuels from Earth into space if safe and secure ways of doing that are realized.

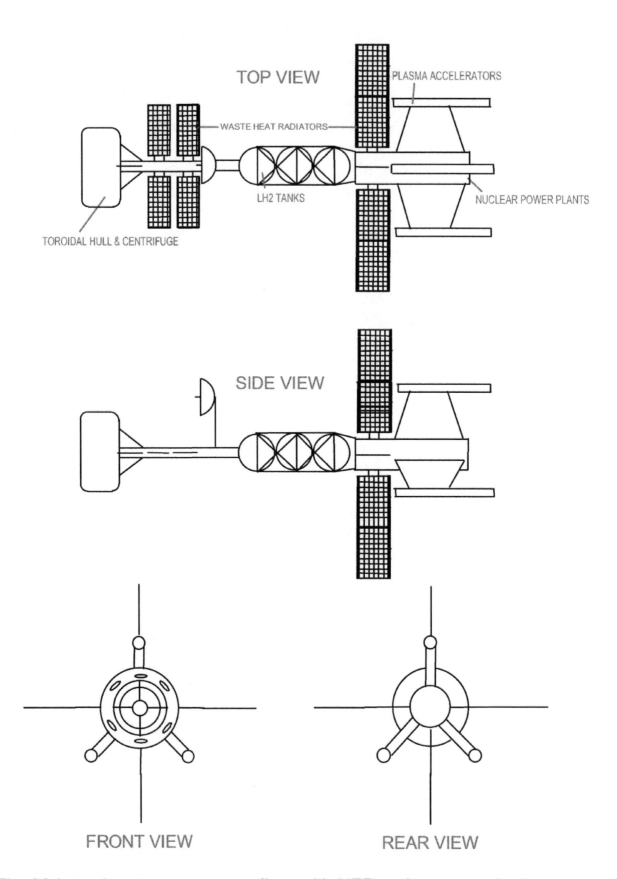

Fig. 33 Interplanetary passenger liner with NEP and superconducting magnetic radiation shield.

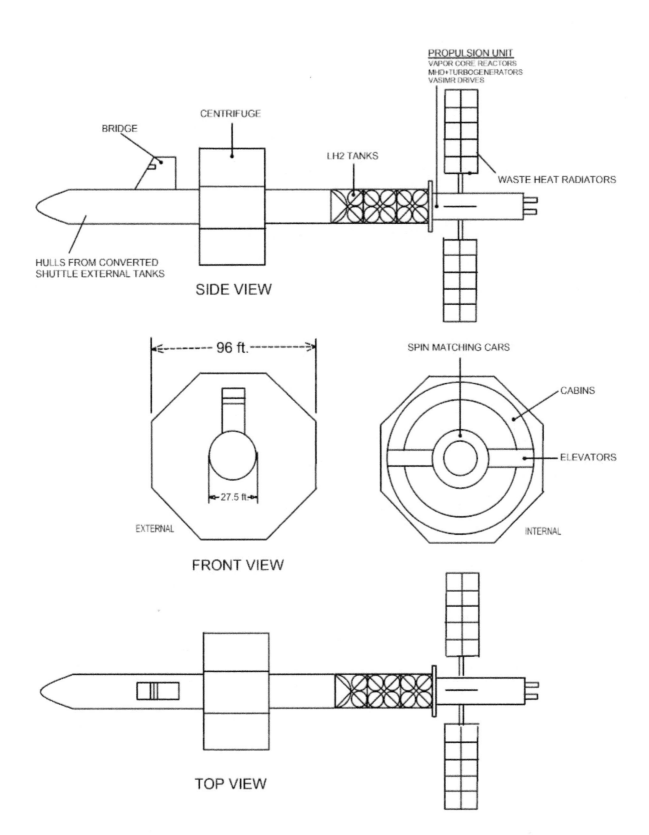

Fig. 34 ET based ship with NEP and a centrifuge.

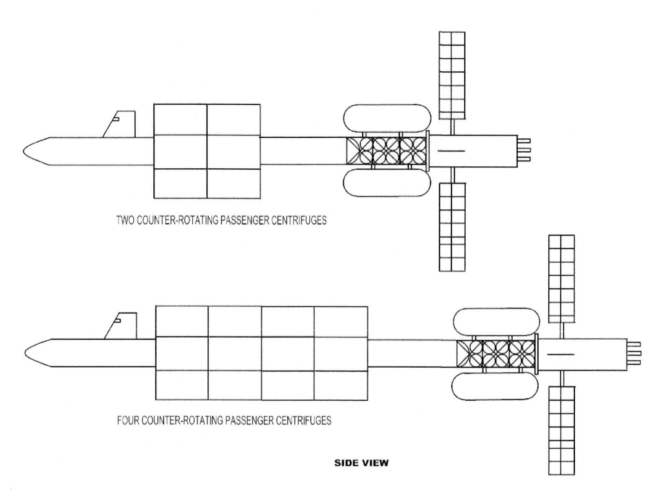

TWO COUNTER-ROTATING PASSENGER CENTRIFUGES

FOUR COUNTER-ROTATING PASSENGER CENTRIFUGES

SIDE VIEW

Fig. 35 NEP ships with two and four centrifuges.

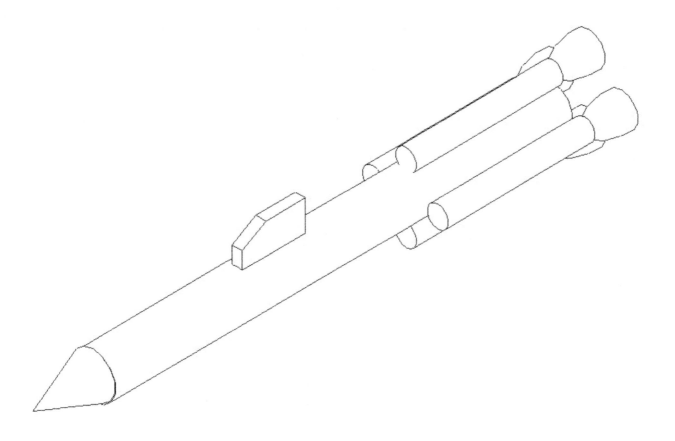

Fig. 36 (above) and Fig. 37 and Fig. 38 (below) CAD drawings by Mark Rode based on designs by David Dietzler

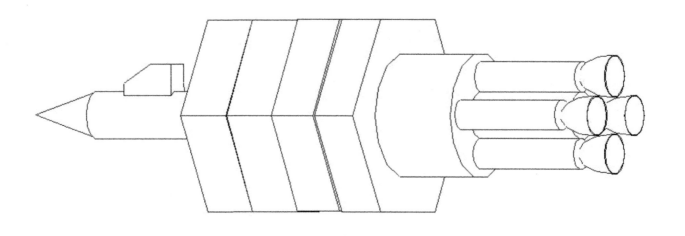

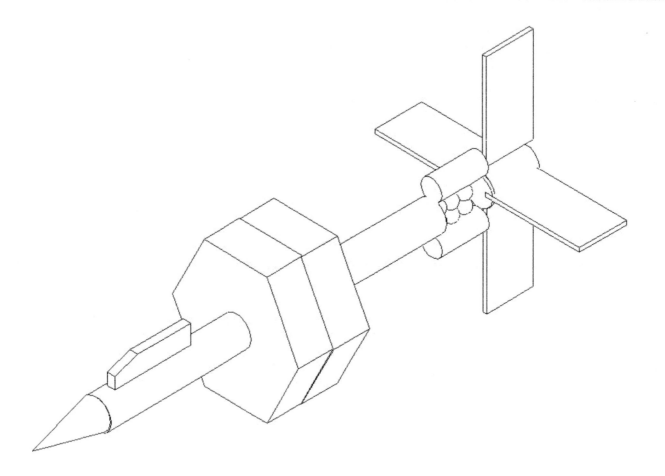

Chapter 16: Law and Order in Space

Interplanetary Government

The popular science fiction trope of united planets governed by a single world government seems unlikely. Look how hard it is just to govern the USA. Governing the whole world would be even harder and in places where people don't like the election results there would be rebellions, insurrections and wars of secession. A single world government could lead to more violence than exists today with over 200 nations on Earth. In the future, there will be numerous nations on Earth, Mars, the Moon, Titan, Mercury and the many moons and asteroids of the solar system.

An interplanetary government would be even more insane. Star empires are probably also nothing but fiction. What kind of God King could rule all the worlds and nations of this solar system with all their different political systems, peoples, languages, religions, customs and cultures as well as other solar systems too?? Even if interstellar travel at high fractions of light speed is developed in the future every solar system will be unique and require local governments. Without law and order and currency too, humans will descend into chaos. Even so, some people prefer anarchy to oppression and denial of liberty. Democracy is the least evil form of government.

There could be something like the UN in space. There could be a United Federation of Worlds. It could be called the Interplanetary United Nations but United Federation of Worlds is an homage to Star Trek's United Federation of Planets. Inhabited planets, dwarf planets, moons, asteroids and free space settlements all qualify as worlds in their own right. There might be federations of space settlements in the solar system that form single nations. Multiple asteroid settlements might form nations too. The UFW would try to resolve disputes between member nations pertaining to space travel related matters. It would not govern affairs within member nations but focus entirely on interplanetary law related to space travel. It would also maintain Space Fleet Command and Space Fleet. Space Fleet could conduct rescue missions, fight smugglers and kidnappers, deliver supplies to remote research stations, transport delegates from various worlds to UFW headquarters and explore and map dwarf planets, asteroids and moons. Space Fleet would consist of dedicated ships and privateers hired to deliver cargoes to other worlds as well as Space Fleet Command--a huge space station and dock in solar orbit; maybe at a Lagrange point for Earth or Mars and the Sun. Space Fleet ships would patrol the solar system ready to perform rescue missions and other tasks.

Space Fleet would be paid for with dues from UFW member nations. We can't expect every nation on Earth, Mars or elsewhere to join the UFW. Some nations may have no interest or financial investment in space travel or just no interest in paying dues to be a member. Individual nations could set their own tariffs and excise taxes and negotiate their own trade agreements with other nations on their planet and on other planets or worlds. Delegates to the UFW will try to reach democratic agreements regarding interplanetary law including safety regulations for spaceships and requirements to assist other ships in distress, sort of like the law of the sea on Earth. Every member nation could send representatives in numbers depending on the population of that nation. Delegates from some worlds might vote as a bloc at times and large worlds like the Earth could have too much power. To even things out a bicameral legislature with a House and Senate could be created. Call it the UFW Assembly. Every member nation could send two senators to the UFW so that all nations have more equal representation when it comes down to matters relating to space travel, trade, exploration, rescue missions, police work and possibly even wars. Bills would have to be approved by both House and Senate. A president, or ersatz king, with power to sign or veto bills will not be desirable. There could be an executive council of secretaries or judges who decide the fate of bills, but would that be necessary? How would the members of the council be appointed? Representatives and senators could be appointed by the governing bodies of different nations. The executive council could be populated by voting in the UFW Assembly if it was even deemed necessary.

There will be an elaborate solar system wide telecommunication system that links all worlds together and enables communication between UFW headquarters where the UFW Assembly is located, Space Fleet Command, all the Space Fleet ships in space, Space Fleet bases and all the nations of all the worlds. The UFW could maintain an organization similar to Interpol on Earth— the Interplanetary Police Network. Information about criminals, smugglers, human traffickers and suspicious characters will be shared so that local police forces can apprehend them. Illegal immigration will probably be next to impossible. Crossing the Rio Grande and hiking through the desert is one thing, but getting a space ship ticket and a passport to a nation on another world and going through multiple checkpoints at space stations where travelers transfer to and from interplanetary spacecraft and shuttles that travel between the ground and low orbit will be very difficult. Anybody rich enough to travel to other worlds probably won't need to enter another nation in space illegally, but there might be asylum seekers and defectors. Refugees from war zones might choose to travel to another world entirely to escape from violence.

Interplanetary war is not impossible. It would probably be very costly to send troops, tanks and fighter jets to another planet to attack a nation below. There would really be no reason for interplanetary defense except for asteroid re-direction systems. A nation on Mars might direct a small asteroid to crash into the capitol city of a nation on Earth instead of landing troops and weapons. Pushing asteroids around and building space weapons carriers could be in violation of interplanetary law. Space Fleet could put a stop to any nation's space warships building program. That would require Space Fleet ships with armaments. Without weapons, UFW ships would be powerless to enforce the law. Space Fleet could also defend worlds from hazardous asteroids and comets with mass drivers and closely monitor asteroid miners so that nobody tries to create a Tunguska level explosion or worse over another world by intention or due to negligence.

The future promises a solar system wide human civilization on planets, moons, asteroids and in O'Neill type space settlements. Since some asteroids and Kuiper Belt objects are fairly large and spherical it seems natural to think of them as minor planets or dwarf planets. Many moons are planet or minor planet sized and spherical and the only thing different about them is that they orbit a major planet. Since there will be settlements on asteroids that are just miles in diameter and those settlements will have centrifuges for a healthy dose of "gravity" and there will be large space settlements that rotate to produce 1 G we have to think beyond planets. These asteroid and space settlements will be like islands and individually or in groups they will form independent nations like island nations in Earth's oceans. Each settlement could have a local democratic government and groups of them could unite under a central national government for mutual assistance and defense. Defense? Yes, there could be conflicts over territory and mining rights, piracy, kidnapping, trade wars and even family feuds. Subsequently, I don't envision a United Federation of Planets but a United Federation of Worlds since moons, inhabited asteroids and space settlements qualify as worlds but they are definitely not planets.

The UFW would oversee Space Fleet. There would be a large space station or settlement where Space Fleet headquarters and Space Fleet Academy are located. Space Fleet's main mission will be rescue and emergency assistance in space. Space Fleet ships will travel in solar orbit and they might have powerful fusion drives to change orbits for rendezvous with ships in trouble. These ships will be as large as the ships of Star Trek. They will have hospitals on board and lots of extra bunk space for survivors of spaceship accidents or malfunctions. They will have spare parts and teams of technicians who can repair disabled spaceships. Space Fleet will also participate in exploration and scientific research in the solar system. They will also fight pirates, smugglers and radical groups if there are any. They might have laser weapons. They could have missiles and automatic cannons too. It should also be possible to board other ships. If Space Fleet troopers can't get through the normal airlocks they could put a large tube up against the hull of the ship and torch their way in. They will need personal arms from cross bows to firearms designed to work in vacuum and extreme temperatures and they will have to be experts in hand to hand combat. There could be zero gravity karate training facilities at Space Fleet Academy.

As time goes by and the population in space grows many nations and settlements may form other governing organizations besides the UFW. Why should nations of multiple space settlements that don't trade much or at all with Earth, Mars and Titan be part of the UFW if they don't want to be? Out in the Kuiper Belt and Oort Cloud they might have no connection with the inner solar system. Fact is, like death, taxes are the only thing we can be certain of...and the only way to support the UFW and Space Fleet is to pay taxes which some nations may want no part of so they don't participate. Perhaps they can rescue their own ships and fight their own pirates and smugglers and do their own scientific research and exploration. Nations that have very few space travelers might not want to pay taxes to the UFW. Perhaps tax support will come from duties imposed on space traveler's fares and space imports and exports. Somehow the UFW will have to enforce taxation and that is the curse of having governments. Anarchy would be even more cursed. The UFW might engage in mining activities to support itself and Space Fleet. There are plenty of metallic asteroids rich in precious metals in space. Platinum, iridium, palladium, rhodium, gold, copper from metallic asteroids and hydrocarbon compounds from carbonaceous asteroids might become widely traded commodities. Currency exchanges will be needed or some kind of crypto-currency like today's Bit-Coin might be used. There could be groups of nations that share a common currency like the European Union does today.

The solar system will be an endlessly fascinating place rich with diversity. There will be cities and nations on other planets, asteroids and in free space to visit and explore. Eventually there will be interstellar settlement and other solar system wide civilizations beyond imagination.

A UFW Settlement on Mercury

Negotiation, deliberation and voting by radio would be difficult given the light speed time delay but not impossible. Meetings of the UFW Assembly would involve delays of minutes for delegates of the inner solar system and hours for those in the outer solar system. There may be times when the need arises for a gathering place and real time face to face meetings. But where to put UFW headquarters???

Mercury presents an interesting possibility. Due to Mercury's short orbital period launch windows to and from every planet in the solar system open up more frequently than on any other world of the solar system. To or from Mercury to or from Venus there is a launch window every 145 days. For Earth every 117 days. Mars and Mercury come within range every 101 days. Launch windows for Jupiter and Saturn are available every 90 and 89 days respectively. Locating UFW HQ and the Assembly on Mercury would ease travel and scheduling challenges for diplomats and their entourage. By comparison, launch windows for Earth and Mars open up about every two years. Launch windows for Earth, Jupiter and Saturn every year. For Mars to or from Jupiter and Saturn about every two years. It will take perhaps six months to travel to or from Mercury to or from Saturn's mons with NEP ships. Travel between Saturn and Earth or Mars will take somewhat less time but travelers would have to wait up to two years for a launch window.

The only caveat is proximity to solar storms. The settlement on Mercury could be safe because it would be built mostly underground beneath many meters of rock and regolith. It would probably be located in the polar regions where temperatures are less extreme and ice has been detected in permanently shadowed craters similar to those that exist on the Moon. The thin hulls of ships will not offer much protection from solar particle events. Ships will need powerful but low mass superconducting magnetic shields and solar flare shelters.

While diplomats and their assistants might not live full time on Mercury but go back home to their worlds when the Assembly was not in session, there will have to be full time staff that lives on Mercury to keep everything functioning including the farms and manufacturing shops. To have a functioning society and not a remote DEW line radar base sort of place there must be farms for fresh food, housing for workers and their families, hospitals, schools, churches, recreation centers, stores, theaters, hotels, police stations, courts of law, jails, factories, etc. Supply ships will also come and go so there will have to be people on Mercury to process all the cargo. The cargo ships will be fully or partially robotic. There will have to be underground warehouses where humans and robots deal with all the freight. Solar energy will be abundant so there really would be no need for imported uranium or helium 3, unless these nuclear fuels are needed for spaceships.

Chapter 17: Humans Into Space

Robots and Humans

Robots will be used extensively for bootstrapping Moon bases and building solar power satellites and other structures in orbit. Some robots will have autonomous AI control and others will be teleoperated by humans in habitat on the Moon and in space, or by controllers on Earth. Teleoperated robots on the lunar surface will be somewhat difficult to control on Earth because of the nearly three second delay due to the speed of electromagnetic waves and the distance between Earth and Moon. Robots in GEO will present a fraction of a second delay, but it will be possible to operate them from Earth. Humans on the Moon and in GEO space stations will be able to operate robots without this hindrance. If and when robots encounter obstacles or break down humans will be able to suit up and go out into the vacuum and fix them.

Earlier, we discussed the mass and costs of moving 1000 tons to the lunar surface with tugs using solar electric propulsion (SEP) and landers with space storable hypergolic propellants. This would all be done robotically. Future robots might not need human supervision, but let's say they do. How do we transport humans to the Moon, GEO stations and other places and keep them alive?? What will this cost??

Moving Humans

Part of the 1000 tons sent to the Moon consists of inflatable habitat and supplies. Robots will mine regolith, extract oxygen, inflate the habitat, and cover them with regolith for radiation shielding prior to human arrival. They will also store up liquid oxygen for manned landers. Humans will not travel to the Moon or GEO in ships with SEP. That would take too long and expose them to radiation in the Van Allen Belts. Reusable chemically propelled rocket vehicles refueled in LEO at orbital depots will be needed. Falcon Heavy rockets could put the modules for the depots in orbit where they are assembled telerobotically. They could also shuttle up 50+ tons of propellant with each launch to stock up the depots. A Falcon 9 or similar rocket could send a Dragon capsule with seven workers on board to LEO for about $70 million. The capsule would rendezvous with the LEO space station/fuel depot where the workers would transfer to a two stage vehicle designed to reach low lunar orbit (LLO). One stage will be capable of descending to the lunar surface. It will land with enough fuel for return ascent to LLO and reload with LOX on the lunar surface. The other stage will remain in lunar orbit until the workers are ready to return to Earth. It will have an ablative heat shield for aerobraking into LEO with both stages. It will cost the Moon mining company over $10 million per human worker stationed on the Moon, so the number of humans will be minimized by the use of automation.

Workers can be sent to GEO stations for SSPS construction in a similar way. They can rocket up to LEO in Dragons on Falcon 9s. Spacecraft similar to the Moon vehicles will fuel up in low orbit and transfer to GEO. To reach GEO a delta V of about 5,000 mph is needed. The ship will travel in an ellipse that reaches GEO where another rocket burn for a dV of 3000 mph circularizes its orbit. To return to Earth a retro burn of 3000 mph is needed with aerobraking into LEO where the workers transfer back to the Dragon docked at the propellant depot and return to Earth. It actually takes more velocity to travel from LEO to GEO and back than it does to reach LLO. Of course, no fuel and oxidizer are needed for a lander stage. It will still cost millions for each worker sent to high orbit so the number of humans will also be minimized by automation in orbit.

Lunar Tourism

When infrastructure has been built up on the Moon and in space and mass drivers are launching enormous quantities of regolith into space the amount of propellant sent up to orbital depots from Earth will be decreased. Oxygen generated by silicon and metal production will be used to stock up depots with LOX. Excess silicon, iron and calcium will be used to make metallic powder slurry fuels. More advanced launch vehicles like hybrid jet/rocket space planes or big reusable rockets that can put 100 passengers in LEO could get the price to space down to a few hundred thousand dollars per person. Humans aren't all that heavy. One hundred of them will weigh about ten metric tonnes, or about 220 pounds apiece so that should cover luggage. It should also be possible to piggyback 20 or more tonnes of liquid hydrogen up to LEO. This will be converted to silane as discussed earlier.

Eventually sizable passenger liners will be built on the ground in separate sections, rocketed to LEO and assembled in space. Propellant depots will grow in size and in number to supply these ships with fuel and oxidizer. They will haul hundreds of people at a time to LLO. Landers, or "Moon Shuttles" built and fueled on the Moon will transport them to the surface. Industry will have grown large on the Moon and luxurious accommodations will be available. Hotels and resorts will be constructed at places of scenic natural beauty. Space farms will supply delicious cuisine and alcoholic beverages. Entertainment venues of all sorts will exist. With all the food, drink and rocket propellant being produced on the Moon or in space, the cost of traveling from LEO to the Moon and staying there will be much less than what it was when everything had to be rocketed up from Earth.

Space Settlement

The first workers in space will be sent there at company expense. The cost to the Moon mining companies and SSPS builders will be more than all but a few will be able or willing to pay. After substantial industry and development exists in space multimillion dollar vacations in orbit and on the Moon will be offered. In time, the cost will come down and middle class citizens will be able to finance a vacation in space, but it won't be cheap. There will be people who want to do more than just visit an Earth orbital hotel or resort on the Moon. Some will satisfy their wanderlust by finding jobs in space while others will become permanent residents not just on the Moon, but in free space settlements, on Mars and other worlds.

Settlers might save up for years, sell their homes, autos and everything else they can't carry in a trunk or suitcase, borrow money, and head for the high frontier. Some of them might work with hotel and resort owners to establish businesses in these places. Doctors, restauranteurs, bar keepers, boutique owners and even lawyers might set up independent businesses. The hotel and resort owners won't have to take care of every detail of human existence for their guests so they will welcome merchants of all sorts. An entrepreneur could take all his money, move to the Moon, rent out some shop space, pay some manufacturers to make an oven and buy food from the farms, which might be independently owned and operated, and set up a pizzeria. He wouldn't have to haul anything with him besides his wallet and his suitcase. Those with more money to invest might buy vehicles from the mining companies and take people on tours. Some might even own their own small suborbital spaceships and take passengers on rides, deliver mail and parcels, or transport prospectors out into the boondocks. These hearty souls might return to Earth someday with a large bank account and others might choose to stay in space for life.

There could also be people who aren't interested in hotel and resort towns or making money off of rich tourists. They might pool their resources and create their own communities beyond the bounds of Earth and all its rules, regulations and troubles. Such people would probably be more interested in emigrating to Mars rather than the Moon. Mars has more gravity than the Moon so it may be easier to stay healthy there for decades and decades. Weight bearing exercise might become as widely practiced as good hygiene is today. Mars has resources of water in permafrost and in polar ice caps. It has an atmosphere consisting mostly of carbon dioxide with some nitrogen and argon. There is also frozen CO_2 at the poles. The day length is only a half hour longer than Earth's and there are seasons on Mars. Habitat will still have to be covered with several meters of regolith to shield against cosmic rays but crops, which have a higher radiation tolerance than do animals, could be grown in plastic domes on the surface. There might be rich ore deposits formed by hydrological processes, unlike the Moon where everything is homogenized by eons of meteoric bombardment and there has never been any water.

Lunar and Earth orbital industry will probably grow to the point at which large spaceships could be built. Perhaps the chemical rocket powered ships built for lunar tourism could be modified with nuclear electric propulsion systems for rapid transit to Mars. The settlers would have to pool their money and buy a spaceship or rent one from a company that sees a lot of customers who want to go to Mars. Future martians could stake out some land on Mars for free because there's nobody to fight over it. They would be wise to send robotic spacecraft to Mars ahead of time and set up living arrangements for the settlers. At first, they might live in inflatable plastic modules. As their communities grow they will build nice big houses underground with local materials like bricks, glass and metal. Hydrogen from permafrost and polar ices along with carbon from the atmosphere and polar caps will be combined to make plastics. Some nitrogen will be needed to make Kevlar but Mars' atmosphere will provide that. A combination of mechanical and biological processes will be used to make necessities. Food will be grown in plastic domes. Perhaps genetically modified plants that can flourish in the less intense sunlight of Mars and survive extreme cold will be cultivated. There might even be livestock. Vegetarianism is all healthy and wonderful, but most people demand their meat!

The martian settlers will build everything with their own hands, therefore they will own it and not have to pay anybody. It would be a rather libertarian paradise until they realized they needed a town hall and some police forces to resolve domestic disputes and control unruly teenagers. If some wealthy criminals land nearby and try to take over their town they will need guns. Let's hope that doesn't happen and the frontier remains peaceful, but all you have to do is watch the news and you come to expect the worst out of humanity.

It looks like they won't have to pay for much besides their transportation to Mars, robots and other necessary equipment to bootstrap a community, and transportation of that equipment to Mars. If 100 ordinary people each sold everything they own they could each come up with several hundred thousand dollars so together they will have millions. If this is enough to get their bodies and their machines to Mars the rest is free for the taking. Water, carbon, nitrogen, regolith, solar energy, land, and other natural resources will be theirs. No Earthly government will have rights to their claims. Even the communists and Muslim radicals won't want to bother with them. Everything they have will be the product of their own labor be it making plastic domes, operating robots or brick making machines, programming computers, driving bulldozers, or any one of a thousand imaginable tasks on Mars.

Free Space

Many people are not interested in living on the Moon, Mars or any other large celestial body. They want to escape the planetary trap and live in free space settlements like those envisioned by O'Neill. Settlements anywhere from a few hundred meters to several miles in diameter have been imagined. The most massive part of these will not be the metal hull but the cosmic radiation shield consisting of several meters thick layers of regolith and water. As a result, free space settlements would require hundreds of millions of tons of material.

Al Globus has proposed smaller settlements without radiation shields in equatorial low Earth orbit protected from cosmic rays by the geomagnetic field.[94] Subsequently, they could weigh less than previous designs by factors of 100 to 1000. These settlements would not be as far away as EML5 or solar orbit. They could be accessed in less than an hour by rocket planes or other launchers lifting off from equatorial bases. They would still weigh on the order of several thousand tons. Launching the materials for these from Earth seems like it will be prohibitively expensive.

Mining the Moon and creating infrastructure in space for SSPSs will result in a lot of by-product silicon, iron and calcium silicate slag along with oxygen. Much of this could be used for interlunar rocket propellant as discussed earlier. It could also be used for space construction. A fraction of the available silicon and almost all of the aluminum, titanium and magnesium from lunar regolith will be in demand for powersats. Iron, slag and glass made by re-oxidizing silicon or extracting it from an intermediate could be used to build settlements in ELEO. Iron would most probably be converted to steel with carbon hauled up from Earth or mined on the Moon. This would require construction stations in ELEO staffed mostly by robots. The construction stations could be composed of modules and hardware launched into space with Falcon rockets. The fate of the BFR remains to be seen. Falcon 9 and Falcon Heavy exist and we know what their price tags are.

If a rocket based on the Shuttle external tank is built a large number of aluminum ETs could become available in ELEO for raw materials. Eventually NEOs could supply glass, metals, water and hydrocarbons to Earth orbit. So potentially there are multiple resource streams for building free space settlements within Earth's magnetic field. Since the market for SSPSs and materials to build them will dry up once the demand for space energy is met, the Moon miners could still sell materials to ELEO construction projects. With a smaller market for SSPSs to replace retired powersats that will be dismantled and their materials recycled, a large market for propellant and space ships, and a growing market for ELEO settlements, the Moon mines could stay in business for decades.

What about people who want to live in free space beyond Earth orbit? Their best bet might be to travel to a NEO with plenty of supplies and machines, mine it in space and bootstrap up their industry until it was possible to build large rotating settlements with steel hulls and thick mass shields. The first settlement could be just a few hundred meters wide while later settlements could be up to 20 miles long and 4 miles wide for millions of inhabitants. It seems doubtful that the Moon mines could supply enough material to build settlements the size of San Francisco at EML5 or EML4. Asteroids near Earth and in the Main Belt seem like the only realistic sources of material for cities in space. Pushing large asteroids into Earth orbit with mass drivers is not a good idea unless you want to risk mass extinction. Small asteroids could be pushed to Sun-Earth Lagrange points or high orbit but some wouldn't even want to risk that. It seems NEOs will be mined by robots and the materials moved back to Earth-Moon space with spaceships.

Mars settlers will be fortunate enough to have two large asteroids already in orbit-the moonlets Deimos and Phobos. If the martians want to build orbital settlements with materials from these two objects it should be possible for them to do so. If the moonlets don't have much in the way of carbon, nitrogen and hydrogen, mass drivers on Mars atop one of the giant shield volcanoes could supply these elements. Many Mars settlers might feel mining Deimos and Phobos would be acts of desecration. They might allow bases and underground tunneling but they might hate the idea of hogging out giant open pit mines on either one of Mars' natural satellites. Why would Mars need space settlements in orbit? They have an entire planet to settle and the Main Belt isn't that far away!

Asteroid miners could do what Mars settlers did and join up with a community of like minded people, sell everything they own, pool their money, buy up the necessary equipment for carving out a home in space and hire a spaceship to take them to their chosen asteroid. If they are successful they could build larger and larger habitat and sell real estate inside to other settlers and go into the tourism business too. By the time they build settlements for a million people they might sell enough real estate to make hundreds of billions or even trillions of dollars. Free space settlements have their appeal. Normal gravity, life in a pleasant setting, no severe weather, freedom from life at the bottom of a gravity well that makes it easy to travel to other worlds of the solar system, mild temperatures and access to weightlessness whenever it is desired. If they build it, millions of people might actually come.

A Space Faring Civilization

As settlements and populations grow in Earth orbit, on the Moon, in solar orbit, on asteroids, on Mars and the moons of the outer planets, even into the Oort Cloud, human diversity will increase. Old cultures could be transplanted to new worlds and entirely new cultures could evolve in space. The human population could increase and this could lead to an explosion of art, music, literature, invention and scientific discovery. Information will travel between the worlds of the solar system at light speed. Humans and freight will be somewhat slower. Cultural diffusion would take place most probably via radio and laser beams. Despite thousands of languages and hundreds even thousands of worlds, popular music, hit songs, could electrify trillions of people and reap fantastic rewards for the artists and producers. Interplanetary copyright laws will have to be agreed on by civilized worlds.

Travelers will want to tour as many worlds as they can. It might be necessary to live well beyond 75 years to do so. Advances in life science and genetic engineering might make it possible for future humans to live for centuries. That would give them enough time to learn everything they need to know and work to earn and save up the money needed to travel. They could also invest over the centuries and become very wealthy.

Robots might do grunt labor and AI may be put to work on problems humans can't solve, but there will still be a need for human creativity, ingenuity and intuition. We can only hope that a synergy will exist between robots and humans that leads to a space faring civilization with great material wealth for all. If poverty is erased, thanks to technology like robots and the acquisition of solar system wide resources, the crime rate should decline and wars could become less common. Higher living standards throughout the solar system could mean better schools, free college and graduate school, higher pay for teachers who are now some of the most undervalued people in society and smaller classroom sizes with lots of individual attention for students who need it. Raising better kids could be the real key to preventing crime and war. Well behaved and well educated kids, whether their brainpower has been boosted by genetic engineering or not, could even conquer all disease, as well as the galaxy.

A vast amount of human potential is available right now even without genetic improvement, but it is going to waste because of global poverty. Rising living standards will mean opportunity for talented persons and this will bring about even more progress. When genetic tweaking of human beings makes it possible to create post-humans or trans-humans we will become a species capable of going to the stars. People of the future with extended life spans who stay young for centuries will have the time to learn far more than the people of today and they will be able to work in multiple fields of professional endeavor. Many will pursue artistic and athletic careers. Higher living standards and educational opportunities will make it possible for humans to do interesting things while robots do the grunt labor in automated factories. Industry freed people from subsistence farm labor. Robots and genetic science will free people from drudgery.

A space faring civilization that even reaches out to the stars won't be worth living in if old prejudices and injustices follow us into space. Chances are we will never reach the high frontier if we continue to stay mired in the hatred and conflicts resulting from superstition and ignorance on Earth. Higher global living standards achieved partly by tapping vast quantities of cheap energy from outer space will mean educational opportunities that dispel ignorance. Many wars and criminal activities are the results of scarcity. Cheap energy, robots and access to virtually unlimited space resources will put an end to scarcity. The promise of tomorrow makes the explorations and scientific research of today worth every dollar invested.

Chapter 18: Beyond the Moon

From the Moon to Mars

It is not necessary to industrialize the Moon and Earth orbit to go to Mars. Spacecraft can be launched straight to Mars as proposed by Dr. Robert Zubrin with his Mars Direct plan.[95] This would involve developing a heavy lift rocket that can send vehicles on 180 day trajectories to Mars. The first vehicle would be the Earth Return Vehicle (ERV). Once it has landed and manufactured fuel and oxidizer using a small amount of onboard liquid hydrogen and CO_2 from the martian atmosphere the second vehicle or Habitat with four persons on board would be launched. Propellant for retro-rocketing into Mars' orbit would not be necessary. The vehicles would aerobrake directly into the planet's atmosphere, deploy parachutes and fire small rockets just before touchdown. After a 540 day stay on the red planet the explorers would board the ERV loaded with fuel and oxidizer made on Mars and lift off to return to Earth in 180 more days.

This would be a great adventure and there is really no reason to practice on the Moon. If you want to go to Mars you simply go to Mars. Rocketry is not the barrier. Life support systems that can operate for years without resupply from Earth are needed. Volunteer explorers willing to endure great lengths of time in small space vehicles and a small but significant increase in their chances of getting cancer due to space radiation exposure could probably be found, but can they really be kept alive with physio-mechanical life support systems? The ERV and Hab vehicles are too small for a farm and Closed Ecological Life Support System (CELSS). Once on Mars oxygen can be extracted from CO_2 in the atmosphere or water from the regolith. Life support challenges might be surmounted.

I don't know if Mars Direct missions will ever happen. The Space Launch System rocket would probably be large enough to do the job and the US government is aiming at the Moon presently. Unlike Moon mining and SSPS building there aren't any foreseeable profits to be made by Mars missions. Mars Direct missions could return a wealth of data about the red planet the way the Apollo missions did for the Moon. This could be of immense value to future space settlers.

If Moon miners and powersat builders create infrastructure in space, it could become possible for large ships to be built in space from lunar materials and NEO resources. Metals from the Moon launched by mass driver could be delivered to space shipyards for just a few dollars per kilogram. These ships would be powered by chemical and nuclear electric propulsion systems with propellants from the Moon and NEOs to reach Mars in about six weeks. Rapid passage to Mars would alleviate the life support challenge. It would be much easier to store dried food, recycle water, store oxygen and use lithium hydroxide cannisters to remove CO_2 from the air for 39 days than it would be for 180 days to several years without resupply. This shorter transit would greatly reduce radiation exposure for the settlers. If the ships have superconducting magnetic radiation shields even less radiation exposure would be endured by travelers. Mars settlers might travel in ships with centrifuges to supply partial gravity that makes life more comfortable during the passage.

Robotic ships would precede humans and establish well shielded dwellings and possibly even space farms. Thousands of settlers might travel to Mars and embark upon the task of building power plants and greenhouse gas factories to terraform the planet. They would probably have to devote their entire life's fortunes to this enterprise. Eventually they might earn money by selling chlorine, copper and zinc to lunar and Earth orbiting industries and they might even build hotels on Mars and accept tourists. While the first explorers might carry out missions equivalent to a dog sled to the South Pole and the early settlers later on could manage to live in spartan accommodations, industrial development on Mars could make nice spacious homes a reality for later settlers and tourists. It would take all kinds of people to make real growing communities on Mars that people would want to raise children in. Everyone from convenience store clerks, Italian chefs and delivery drivers to security guards, business administrators, school teachers, doctors and engineers would be needed. The percentage of people working in the greenhouse gas factories might actually be rather small. Most of the settlers will be devoted to building communities worth living in. They will have to provide everything from diapers to religious services. There would definitely be opportunities for anyone skilled enough to produce a good or service and sell it to the settlers be it pizzas or music lessons for kids.

The Main Belt

Not everyone has a passion for Mars. Some people want to build their own worlds in free space. Robots and humans could bootstrap up industry on or inside asteroids and create the infrastructure needed to build rotating metal bubbles and cylinders with dimensions measured in miles. Even the largest asteroid in the Main Belt, or dwarf planet, Ceres, doesn't have enough gravity for healthy living. There are many small worlds in the Main Belt that would be interesting to explore but they don't offer much hope for long term inhabitation. Even so, the Main Belt would be an interesting place to live inside a large space habitat rotating fast enough to produce Earth normal gravity. Large solar collectors would be needed to harvest energy that far from the Sun, but there don't seem to be any reasons that couldn't be done. Small asteroids just a few kilometers in diameter would provide billions of tons of rock, metal, oxygen, hydrocarbons and water from ice. Perhaps a trillion people could find a home in Main Belt space settlements. There are also so called Trojan asteroids at Sun-Jupiter Lagrange regions 4 and 5 and many asteroids within the orbit of Mars.

Bootstrapping on an asteroid would be a lot like bootstrapping on the Moon, but with less gravity to struggle with. Construction of large habitat would be a job that architects, engineers and space construction workers would have already mastered in Earth orbit. Everything the settlers create with their own hands and their robot servants will be theirs and they won't owe anyone property taxes or be forced to pay for licenses. The only thing they will need money for is equipment and transportation. They might borrow money for this and pay it back by producing some goods of value to people back on Earth or the Moon. Platinum group metals come to mind. Robotic prospecting surveys conducted by governments or private enterprises might locate valuable resources, but that won't give them the right to make any claims. They would sell the information to settlers and let them go out and brave the dangers of space. Territorial claims will be made no matter what unrealistic utopian socialist lawmakers try to force on people. That's just human nature. Whoever digs it up owns it.

Small groups of families might eventually grow into huge nations of space dwellers. Eventually, their own economy will grow and they will need currency. It's hard to imagine a barter economy for an advanced technological civilization. Precious metals sold to nations on other worlds will make it possible for the settlers or Main Belt nationals to buy goods and services from those nations on the Moon or Mars that they can't produce themselves. A thriving economy between the Main Belters, Mars, the Moon, Earth and perhaps some other worlds could emerge. Travel and tourism would not be discouraged, except perhaps by some of the stricter Muslim settlements that don't want cultural contamination by foreigners. Large numbers of travelers, be they merchants or sightseers, would need basic commodities like oxygen, water, food and propellant. They would have to trade for these things. Precious asteroidal metal mining might be exceeded by the space travel industry as a source of revenue. Hydrogen propellant and uranium will be hot commodities.

Life in free space settlements with Earth normal gravity will not require weight-lifting as common as taking a shower as it would on the Moon and Mars. It's always possible some drug will be developed that prevents bone and muscular atrophy or the martians and lunans will have their off-spring genetically engineered for adaptation to life in low G. Learning to use a spacesuit and get around in weightlessness or low gravity environments will be as common amongst young lunans, martians and Main Belters as learning to swim and drive is today on Earth in developed nations. Opportunities for exploration and adventure on the larger asteroids will be numerous. There could be underground habitations complete with farms in many of these asteroids for temporary residents. What would it be like to stand on Ceres or Juno? What fantastic geological features exist out there? Certainly, there will be John Muirs and Ansel Adams'es of outer space who explore and photograph vistas on the Moon, Mars and dozens of asteroids. Nature's beauty is an irresistible force.

To the Outer Solar System

With the success of settlements on the Moon, Earth orbit, Mars and the Main Belt, humans will confidently move out to the moons of the outer planets Jupiter, Saturn, Uranus and Neptune. What would it be like to stand on one these natural satellites like Callisto, Titan, Oberon or Triton and look up at a Gas Giant planet in the sky amongst the stars?? The lure of the unknown, the call of the wild, the urge to explore cannot be stifled.

There could be financial gains involved. While many asteroids contain water and hydrocarbons, those asteroids have to be excavated, their materials crushed into powder and roasted to drive out the water and hydrocarbons. Titan has an atmosphere of nitrogen and methane and the probability of hydrocarbon lakes. Could it be easier or more economical to pump up liquid hydrocarbons on Titan and rocket them into space with jet-atomic vehicles and load up robotic space tankers that haul the stuff anywhere in the solar system where there are people who need carbon for steel and composites and hydrocarbons and nitrogen for plastics, glues, resins, paints, chemicals, drugs, fungicides, and so many other products now derived from oil on Earth? There might be intense competition with asteroid miners and ice miners on other moons, but Titan might become the Persian Gulf of the solar system. Titan dwellers might convert the hydrocarbons to thousands of finished products and just ship them to settlements all over the solar system instead of tanking crude which won't be of much use without an industrial base to process it. There isn't much solar energy out on Titan and we don't know if there is any potential for wind or geothermal energy, but there is always nuclear power.

One of the most interesting things about Titan is that it has an atmosphere. It will be possible to fly airplanes there. Aircraft might have electric motors or use turbines that draw in methane from the atmosphere and burn it with LOX carried onboard. The LOX would come from rocks and regolith. Balloons and dirigibles filled with hydrogen could be used safely in the non-oxygenated atmosphere. Titan may turn out to be one of the most interesting and lucrative worlds of the solar system.

To the Stars!

It all began with Moon mining and it won't end with the solar system. Free space habitats or settlements of size so great life within them is indistinguishable from life on a planetary surface with day and night, mild breezes, clouds and rain could be built all over the solar system; even out as far as the Edgeworth-Kuiper Belt and the Oort Cloud. Settlements out there could consist of metal hulls with cosmic ray shields made of ice. They could be illuminated within by microwave sulfur lamps that mimic the spectrum of the Sun without the infrared and ultraviolet light. Sunlight will be dim that far out, but controlled nuclear fusion might finally be perfected by the time humanity reaches this far out into space. There's plenty of deuterium in cometary ices for fusion fuel. Settler populations could expand into the outer reaches of our Oort Cloud over the centuries or millenia and cross over to the Oort Cloud of the Proxima-Alpha Centauri system.

Many have imagined interstellar flight with generation ships, fusion drives or anti-matter propulsion. Generation ships would travel for centuries to the nearest stars. The ships would have to be as massive as free space settlements outside of the geomagnetic field. That means they would require multi-million ton cosmic ray mass shields. Not only are these ships too slow, they are too heavy, and they require closed ecological life support systems that can operate without inputs or outputs for centuries. Who would want to start out on a journey and die enroute with only the expectation that a remote descendent would see the new solar system? While the generation ship was lumbering out to another star system new technology might be created back home that made rapid interstellar travel possible...so why bother???

Fusion drives might make interstellar travel possible in mere decades. Still, for people who only live an average of 75 years this would be too long. If lifespans are extended to several centuries through genetic engineering a journey lasting 40 years might be worthwhile if living conditions on the ship are good enough. If suspended animation is possible even better. Not only would suspended animation prevent boredom on long voyages, it should stop the aging process and reduce life support demands. A smaller faster ship might be possible since all kinds of recreation facilities would not be needed and neither would farms and a closed ecological life support system to provide food and deal with human wastes.

The most interesting fusion propulsion system is the Bussard ram jet. This would not require vast amounts of reaction mass. A magnetic scoop would collect interstellar hydrogen, compress it in a magnetic field, and create a powerful exhaust jet. Objections to the Bussard ram jet are numerous. It has been said that compressing normal hydrogen in a magnetic fusion reactor would be too difficult to ever be possible. The magnetic scoop would collect so little hydrogen and experience so much resistance that the ship just won't work. Perhaps antimatter could be used to energize the hydrogen. It might even catalyze fusion for an enhanced effect. Perhaps the antimatter energized exhaust could overcome the resistance of the magnetic field with interstellar plasma. The magnetic field could also be used as a brake to slow the ship down. Zubrin and Andrews have proposed a magnetic sail for braking in interstellar space.[96] The magnetic sail could also ride solar winds closer in to stars and allow ships to navigate in distant star systems without using any propellant. Magnetic sails might be a great way to propel freighters and tankers in our solar system.

A final method of propulsion would be laser sails or magnetic sails propelled by particle beams. Enormous solar energy complexes at Sun Venus Lagrange point 1 perhaps to power lasers and light sails or particle accelerators of gigantic size could propel ships up to high fractions of light speed. Magnetic sails could be used to brake these ships. If we combine rapid journeys lasting but years with extended life spans and suspended animation, interstellar voyaging is very appealing. It isn't warp drive or travel through wormholes, but it would be good enough. As for warp drives and wormholes, who knows what discoveries in higher physics will lead us to in the future? One thing is certain, the solar system is so big it could keep explorers and settlers busy for thousands of years.

References

1) Mr. John C. Mankins. "SPS-ALPHA: The First Practical Solar Power Satellite via Arbitrarily Large Phased Array" (A 2011-2012 NASA NIAC Phase 1 Project)   15 September 2012 https://space.nss.org/media/SPS-Alpha-The-First-Practical-Solar-Power-Satellite-2012-Mankins.pdf

2) Richard D. Johnson and Charles Holbrow editors. "Space Settlements: A Design Study." NASA SP-413. Chapter 5. Appendix I. Scientific and Technical Information Office. NASA. 1977. https://space.nss.org/settlement/nasa/75SummerStudy/Table_of_Contents1.html

3) Franklin R. Chang-Diaz. "The VASIMR Rocket." Scientific American. Vol. 283 No. 5, Pgs.92-93. November 2000.

4) Andrew V. Iln et al. "VASIMR Human Mission to Mars." Space, Propulsion & Energy Sciences International Forum March 15-17, 2011, University of Maryland, College Park http://www.adastrarocket.com/Andrew-SPESIF-2011.pdf

5) https://www.nasa.gov/centers/kennedy/about/information/shuttle_faq.html#1

6) T. A. Heppenheimer. Colonies in Space. Chp. 6-The Moon Miners. 1977, 2007. https://space.nss.org/colonies-in-space-chapter-6-the-moon-miners/

7) Dr. Peter J. Schubert. Energy Resources Beyond Earth-SSP from ISRU. November 2014. https://ildwg.files.wordpress.com/2015/03/energy-resources-beyond-earth-ssp-from-isru-schubert.pdf

8) Advanced Automation for Space Missions. Edited by Robert A. Freitas, Jr and William P. Gilbreath. Pg. 290. Santa Clara, California. 1980. https://space.nss.org/media/1982-Self-Replicating-Lunar-Factory.pdf

9) Dr. William Agosto. "Lunar Beneficiation." http://www.nss.org/settlement/nasa/spaceresvol3/lunarben1b.htm

10) Matthew E. Gajda et al. "A Lunar Volatiles Miner." nasa-academy.org/soffen/travelgrant/gadja.pdf

11) Decomposition of Carbon Dioxide. http://carbon.atomistry.com/decomposition_carbon_dioxide.html

12) Boudouard Reaction. https://en.wikipedia.org/wiki/Boudouard_reaction

13) Al Globus and Joe Strout, Orbital Space Settlement Radiation Shielding, preprint, pg. 14, 2016. http://space.alglobus.net/papers/RadiationPaper.pdf

14) W. David Carrier III et. al., Lunar Sourcebook, Chp. 9, "Physical Properties of the Lunar Surface," pg. 484, <http://www.lpi.usra.edu/publications/books/lunar_sourcebook/pdf/Chapter09.pdf >

15) Richard E. Gertsch, A Baseline Lunar Mine, NASA SP-509, vol. 3, 1992
http://www.nss.org/settlement/nasa/spaceresvol3/ablm1.htm

16) Advanced Automation for Space Missions. Chp. 4.2.2 Table 4.16.- Lunar Factory
Applications Of Processed Basalt
http://en.wikisource.org/wiki/Advanced_Automation_for_Space_Missions/Chapter_4.2.2

17) Larry A. Haskin, Hydrogen and Oxygen From Lunar Polar Water, Or From Lunar Dirt?
Magma Electrolysis, http://meteorites.wustl.edu/abstracts/nvm3_00/a_m00h02.pdf

18) Dr. William Agosto. "Lunar Beneficiation."
http://www.nss.org/settlement/nasa/spaceresvol3/lunarben1b.htm.

19) Rudolf Keller and D.B.Stofesky. "Selective Evaporation of Lunar Oxide Components."
Space Manufacturing 10, pp. 130-135, Abstract: http://ssi.org/ssi-conference-abstracts/space-
manufacturing-10

20) ditto

21) http://smt.sandvik.com/en/products/metal-powder/3d-printing/

22) http://gpiprototype.com/services/metal-3d-printing.html

23) http://www.planetaryresources.com/news/#pri-blog

24) Titanium Alloys-Alpha, Beta and Alpha/Beta.
https://www.azom.com/article.aspx?ArticleID=915

25) http://www.k-mm.com/large-machining/#large_machining_11

26) http://www.hurco.com/en-us/cnc-machine-tools/machining-
centers/vertical/pages/performance.aspx

27) Peter Kokh, "MUS/cle Strategy for Lunar Industrial Diversification," Lunar Reclamation
Society, 1988 < http://www2.moonsociety.org/publications/mmm_papers/muscle_paper.htm

28) David Schrunk et. al. The Moon: Resources, Future Development and Settlement. 2nd
ed. Praxis Publishing Ltd. Chichester, UK, pg. 54, 2008.

29) Dr. Peter Schubert. "How Moonrocks can Save the Earth." 2008 <
http://moonsociety.org/spreadtheword/pdf/UsingMoonRockstoSavetheEarth.pdf >

30) Geoffrey A. Landis , "Materials Refining for Solar Array Production on the Moon,"
NASA/TM—2005-214014 https://info.aiaa.org/tac/SMG/SCTC/Shared%20Documents/Landis-
Lunar_Production-TM-2005-214014.pdf
31) Track (rail transport). https://en.wikipedia.org/wiki/Track_(rail_transport)

32) Peter Kokh et al. "Railroading on the Moon." 1993.
http://strabo.moonsociety.org/mmm/whitepapers/rr_moon.htm

33) Wickman Spacecraft and Propulsion Co,, "Rocket Propellant from Lunar Soil,"
http://www.wickmanspacecraft.com/lsp.html

34) Peter Kokh, "Spinining Up Glass-Glass Composites Technology," Lunar Reclamation
Society, 1987, <
http://www.moonsociety.org/publications/mmm_papers/glass_composites_paper.htm >

35) Solar Power Satellite System Definition Study, Vol 5, Phase 1, Executive Summary, Final
Briefing, Boeing, pg. 113, 1978. < http://www.nss.org/settlement/ssp/library/BoeingCR160374-
1978DEC-Phase1Vol5FinalBriefing-ExecutiveSummary.pdf >

36) Geoffrey A. Landiss, Aluminum Production on the Moon with Existing Silicon Production,
http://www.asi.org/adb/02/13/02/aluminum-production.html

37) Bonnie Cooper. "Reservoir Estimates for the Sulpicious Gallus Region." Engineering,
Construction and Operations in Space IV. 1994. Abstract:
https://cedb.asce.org/CEDBsearch/record.jsp?dockey=0086432

38) Formaldehyde. https://en.wikipedia.org/wiki/Formaldehyde

39) Darrel D. Ebbing and Mark S. Wrighton ed. General Chemistry 14th edition. Pg. 936.
Houghton Mifflin Company, Boston. 1993.

40) Ethane. https://en.wikipedia.org/wiki/Ethane

41) Heat Pipe. https://en.wikipedia.org/wiki/Heat_pipe#Spacecraft

42) Ebbing and Wrighton. General Chemistry 14 ed. Pg. 651. Houghton Mifflin, Boston 1993.

43) Acetone. https://www.britannica.com/science/acetone

44) Acetone. https://en.wikipedia.org/wiki/Acetone#Chemical_intermediate

45) Acetone. https://en.wikipedia.org/wiki/Acetone#Current_method

46) Bakelite. https://en.wikipedia.org/wiki/Bakelite

47) Schuhmann, M. et al. Aliphatic and aromatic hydrocarbons in comet 67P/Churyumov-
Gerasimenko seen by ROSINA. 2019.
https://ui.adsabs.harvard.edu/abs/2019A%26A...630A..31S/abstract

48) Preparation of Benzene - Aromatic compounds in Organic Chemistry
https://byjus.com/chemistry/preparation-of-benzene/

49) Calcium Carbide. https://en.wikipedia.org/wiki/Calcium_carbide

50) Geoffrey A. Landis. "Resource Production on the Moon." 5th Joint Meeting of Space Resources Roundtable/Planetary and Terrestrial Mining Sciences Symposium, Golden CO, June 10-11, 2014. https://ntrs.nasa.gov/archive/nasa/casi.ntrs.nasa.gov/20140017767.pdf

51) Vinyl Chloride. https://en.wikipedia.org/wiki/Vinyl_chloride

52) 1,2-Dichloroethane. https://en.wikipedia.org/wiki/1,2-Dichloroethane

53) Polyvinyl Chloride. https://en.wikipedia.org/wiki/Polyvinyl_chloride

54) William H. Nebergall and Frederic C Schmidt. College Chemistry. Pg. 445. D.C. Heath and Company, Boston 1957.

55) Glycerol. https://en.wikipedia.org/wiki/Glycerol

56) Castor Oil. https://en.wikipedia.org/wiki/Castor_oil

57) Jojoba Oil. https://en.wikipedia.org/wiki/Jojoba_oil#History

58) New Production Process Makes PLA Bioplastic Cheaper and Greener. New Atlas. https://newatlas.com/bioplastic-pla-cheaper-production-process/38498/

59) https://polymerinnovationblog.com/from-corn-to-poly-lactic-acid-pla-fermentation-in-action/

60) https://www.wikihow.com/Make-Bioplastic

61) https://www.wikihow.com/Make-Paper-at-Home

62) https://boltthreads.com/technology/mylo/

63) https://boltthreads.com/technology/microsilk/

64) Bamboo. https://en.wikipedia.org/wiki/Bamboo

65) Bamboo Textiles. https://en.wikipedia.org/wiki/Bamboo_textile#Bamboo_rayon

66) https://www.thoughtco.com/drugs-and-medicine-made-from-plants-608413

67) https://www.diynatural.com/natural-fabric-dyes/

68) https://thegreenhubonline.com/2018/05/16/easy-diy-how-to-dye-fabric-using-natural-vegetable-dyes/

69) Linus Pauling. General Chemistry. Dover Publications, Inc., New York. pg. 644, 1988.

70)) Rudolf Keller and David B. Stofesky. Selective Evaporation of Lunar Oxide Components. Space Manufacturing 10. AIAA and SSI. 1995. Abstract at: https://space.nss.org/space-manufacturing-10/

71) What Are the Proper Concrete Mix Proportions? BN Products.
http://www.bnproducts.com/blog/what-are-the-proper-concrete-mix-proportions/

72) https://www.engineeringenotes.com/concrete-technology/sulphur-concrete/sulphur-concrete-production-properties-and-advantages-concrete-technology/31868

73) Linus Pauling. General Chemistry. Dover Publications, Inc., New York. pg. 636, 1988.

74) Carbonaceous chondrite. https://en.wikipedia.org/wiki/Carbonaceous_chondrite

75) Stanley Schmidt and Robert Zubrin editors. Islands in the Sky. Chp. 12 The Magnetic Sail. Pg. 199. John Wiley and Sons. New York, 1996.

76) Roger R. Bate et al. Fundamentals of Astrodynamics. Pg 329. Fig. 7.3-1. Dover Publications Inc. New York, 1971.

77) Bryce L. Meyer. Space Farm, Terraforming, and Space Habitats Page
Space Farms, Terraforming, Closed Cycle Farming and Zero-g/microgravity farming.
http://www.combat-fishing.com/animationspace/terraform.html#closecycleeco101

78) T.A. Heppenheimer. Colonies in Space. Chp. 5. Stackpole, 1977, 2007
https://space.nss.org/colonies-in-space-chapter-5-first-of-the-great-ships/

79) http://www.ibtimes.com/new-class-easily-retrievable-asteroids-could-be-captured-rocket-technology-found-1382529

80) Wickman Spacecraft and Propulsion Company. http://wickmanspacecraft.com/lsp/

81) Fraser Cain. "How Far is Mars from Earth?" Universe Today. 2013.
https://www.universetoday.com/14824/distance-from-earth-to-mars/

82) New Scientist. "How to Call the International Space Station." 2019.
https://www.newscientist.com/article/2218359-how-to-call-the-international-space-station/

83) Microsoft. "How much bandwidth does Skype need?" 2020.
https://support.skype.com/en/faq/FA1417/how-much-bandwidth-does-skype-need

84) ATT. Internet Data Calculator. 2020. https://www.att.com/esupport/data-calculator/index.jsp

85) Peter Kokh. A Pioneer's Guide to Living on the Moon. Chp. 6. Luna City Press. 2018.

86) https://www.watercalculator.org/footprint/indoor-water-use-at-home/

87) Bryce L. Meyer. Space Farm, Terraforming, and Space Habitats Page
Space Farms, Terraforming, Closed Cycle Farming and Zero-g/microgravity farming.
http://www.combat-fishing.com/animationspace/terraform.html#closecycleeco101

88) Mark Robinson. 100 Grams of Uranium Equals 290 Tons of Coal. R&D Engineering, Kelso, WA. 1987.

89) Richard Garwin and Georges Charpak. Megawatts and Megatons: The Future of Nuclear Power and Nuclear Weapons. Pg. 150. University of Chicago Press, Chicago, 2000.

90) https://www.nextbigfuture.com/2009/12/vasimr-and-stronger-nuclear-reactors.html

91) "Fusion Energy Gets Ready to Shine—Finally." Wired. April 21, 2020. https://www.wired.com/story/fusion-energy-iter-reactor-ready-to-shine/

92) Daniel Jassby. "ITER is a showcase … for the drawbacks of fusion energy" Bulletin of the Atomic Scientists. February 14, 2018. https://thebulletin.org/2018/02/iter-is-a-showcase-for-the-drawbacks-of-fusion-energy/

93) https://www.energy.gov/supercritical-co2-tech-team

94) Al Globus and Tom Marotta. The High Frontier: An Easier Way. 2018.

95) Robert Zubrin and Richard Wagner. The Case for Mars. Simon & Schuster. New York, 1996.

96) Stanley Schmidt and Robert Zubrin editors. Islands in the Sky. "The Magnetic Sail" by Robert Zubrin. Pg.199. John Wiley & Sons, New York, 1996.

Printed in Great Britain
by Amazon

18076724R00088